Image Processing and Acquisition using Python

T0253468

Chapman & Hall/CRC
The Python Series

About the Series

Python has been ranked as the most popular programming language, and it is widely used in education and industry. This book series will offer a wide range of books on Python for students and professionals. Titles in the series will help users learn the language at an introductory and advanced level, and explore its many applications in data science, AI, and machine learning. Series titles can also be supplemented with Jupyter notebooks.

Image Processing and Acquisition using Python, Second Edition

Ravishankar Chityala, Sridevi Pudipeddi

For more information about this series please visit: https://www.crcpress.com/Chapman--HallCRC/book-series/PYTH

Image Processing and Acquisition using Python

Second Edition

Ravishankar Chityala
IonPath and University of California Santa Cruz

Sridevi Pudipeddi
University of California, Berkeley

CRC Press
Taylor & Francis Group
Boca Raton London New York

CRC Press is an imprint of the
Taylor & Francis Group, an **informa** business

A CHAPMAN & HALL BOOK

First edition published 2021
by CRC Press
6000 Broken Sound Parkway NW, Suite 300, Boca Raton, FL 33487-2742

and by CRC Press
2 Park Square, Milton Park, Abingdon, Oxon, OX14 4RN

First issued in paperback 2022

Visit the Taylor & Francis Web site at
http://www.taylorandfrancis.com

and the CRC Press Web site at
http://www.crcpress.com

Library of Congress Cataloging-in-Publication Data
Names: Chityala, Ravishankar, author. | Pudipeddi, Sridevi, author.
Title: Image processing and acquisition using Python / Ravishankar Chityala
    and Sridevi Pudipeddi.
Description: Second edition. | Boca Raton : Chapman & Hall/CRC Press, 2020.
    | Series: Chapman & Hall/CRC the Python series | Includes
    bibliographical references and index.
Identifiers: LCCN 2020015556 | ISBN 9780367198084 (hardback) | ISBN
    9780429243370 (ebook)
Subjects: LCSH: Image processing. | Python (Computer program language)
Classification: LCC TA1637 .C486 2020 | DDC 006.6/63--dc23
LC record available at https://lccn.loc.gov/2020015556

ISBN 13: 978-0-367-53157-7 (pbk)
ISBN 13: 978-0-367-19808-4 (hbk)
ISBN 13: 978-0-429-24337-0 (ebk)

DOI: 10.1201/9780429243370

To our parents

Contents

Foreword

I first met one of the authors, Dr. Ravishankar (Ravi) Chityala, in 2006 when he was a PhD student at the Toshiba Stroke Research Center, SUNY-Buffalo. Ravi's PhD work in medical imaging was fruitful and influential, and I have been following his post-PhD career ever since. In reading this book, I was impressed by the fact that, despite Ravi's current focus on computing and visualization, his knowledge of medical imaging has only deepened and expanded, which has enabled him, along with his co-author, Dr. Sridevi Pudipeddi, to write a very competent treatment of the subject of medical imaging. Thus, it is a pleasure for me to write a foreword to this very good book.

This is a book that every imaging scientist should have on his or her desk because image acquisition and processing is becoming a standard method for qualifying and quantifying experimental measurements. Moreover, I believe students and researchers need a course or a book to learn both image acquisition and image processing using a single source, and this book, as a well-rounded introduction to both topics, serves that purpose very well. The topics treated are complex, but the authors have done a great job of covering the most commonly used image acquisition modalities, such as x-ray and computed tomography, magnetic resonance imaging, and microscopes, concisely and effectively, providing a handy compendium of the most useful information.

As Confucius said, "I see and I remember, I do and I understand;" this book aims to provide hands-on learning that enables the reader to understand the concepts explained in the book by means of applying the various examples written in the Python code. But do not be discouraged if you have never used Python or any other script language since learning it is very straightforward. As a long-time Perl user, I had no problem installing Python and trying several useful examples from the book. Most of the equations provided in the book are accompanied by codes that can be quickly run and modified for the reader to test new ideas and apply to his or her own research.

Being a medical imaging scientist myself, I really enjoyed reading the sections on x-ray, computed tomography and magnetic resonance imaging. The authors provide a well-balanced introduction to these modalities and cover all the important aspects of image acquisition, as well as image reconstruction and artifacts correction. The authors also provide a large number of references to other books and papers for readers interested in learning more details.

In summary, the strengths of the book are:

1. It teaches image processing using Python, one of the easiest and most powerful programming languages

2. It covers commonly used image acquisition and processing techniques

3. It cements readers' understanding with numerous clear examples.

Alexander Zamyatin
Distinguished Scientist
Toshiba Medical Research Institute USA, Inc.
Vernon Hills, Illinois

Preface

We received feedback from people who bought the first edition of the book and also from experts in the topic while working on the second edition of the book.

We added three new chapters and one new appendix. When the first edition was written, machine learning (ML) and deep learning (DL) were not yet mainstream. Today, problems that cannot be solved by using traditional image processing and computer vision techniques are being solved using ML and DL. So we added one chapter on neural network and another chapter on convolutional neural network (CNN). In these two chapters, we discuss the mathematical underpinnings of these two networks. We also discuss solving these two networks using Keras, a ML / DL library.

We also added a new chapter on affine transformation, a geometric transformation that preserves lines. We also added an appendix on parallel computing using joblib, a Python module that allows distributing tasks to multiple Python process that can run on multiple cores on a given computer.

We added new algorithms to existing chapters and also improved the explanation of the code. Some of the new algorithms introduced are Frangi filter, Contrast Limited Adaptive Histogram Equalization (CLAHE), Local contrast normalization, Chan-Vese segmentation, Gray scale morphology etc.

When the first edition was written, we used Python 2.7 for testing the code. As of January 2020, Python 2.7 is no longer supported. So we modified the code for the latest version of Python 3. We also modified the code for the latest version of numpy, scipy, scikit and OpenCV.

We hope you enjoy learning from the book.

MATLAB® is a registered trademark of The MathWorks, Inc. For product information, please contact:

The MathWorks, Inc.

3 Apple Hill Drive

Natick, MA 01760-2098 USA

Tel: 508-647-7000

Fax: 508-647-7001

E-mail: info@mathworks.com

Web: www.mathworks.com

Preface to the First Edition

Image acquisition and processing have become a standard method for qualifying and quantifying experimental measurements in various Science, Technology, Engineering, and Mathematics (STEM) disciplines. Discoveries have been made possible in medical sciences by advances in diagnostic imaging such as x-ray based computed tomography (CT) and magnetic resonance imaging (MRI). Biological and cellular functions have been revealed with new imaging techniques in light based microscopy. Advancements in material sciences have been aided by electron microscopy. All these examples and many more require knowledge of both the physical methods of obtaining images and the analytical processing methods to understand the science behind the images. Imaging technology continues to advance with new modalities and methods available to students and researchers in STEM disciplines. Thus, a course in image acquisition and processing has broad appeal across the STEM disciplines and is useful for transforming undergraduate and graduate curriculum to better prepare students for their future.

This book covers both image acquisition and image processing. Existing books discuss either image acquisition or image processing, leaving a student to rely on two different books containing different notations and structures to obtain a complete picture. Integration of the two is left to the readers.

During the authors' combined experiences in image processing, we have learned the need for image processing education. We hope this book will provide sufficient background material in both image acquisition and processing.

Audience

The book is intended primarily for advanced undergraduate and graduate students in applied mathematics, scientific computing, medical imaging, cell biology, bioengineering, computer vision, computer science, engineering and related fields, as well as to engineers, professionals from academia, and the industry. The book can be used as a textbook for an advanced undergraduate or graduate course, a summer seminar course, or can be used for self-learning. It serves as a self-contained handbook and provides an overview of the relevant image acquisition techniques and corresponding image processing. The book also contains practice exercises and tips that students can use to remember key information.

Acknowledgments

We are extremely thankful to students, colleagues, and friends who gave valuable input during the process of writing this book. We are thankful to the Minnesota Supercomputing Institute (MSI) at the University of Minnesota. At MSI, Ravi Chityala had discussions with students, staff and faculty on image processing. These discussions helped him recognize the need for a textbook that combines both image processing and acquisition.

We want to specially thank Dr. Nicholas Labello, University of Chicago; Dr. Wei Zhang, University of Minnesota; Dr. Guillermo Marques, University Imaging Center, University of Minnesota; Dr. Greg Metzger, University of Minnesota; Mr. William Hellriegel, University of Minnesota; Dr. Andrew Gustafson, University of Minnesota; Mr. Abhijeet More, Amazon; Mr. Arun Balaji; and Mr. Karthik Bharathwaj for proofreading the manuscript and for providing feedback.

We thank Carl Zeiss Microscopy; Visible Human Project; Siemens AG; Dr. Uma Valeti, University of Minnesota; Dr. Susanta Hui, University of Minnesota; Dr. Robert Jones, University of Minnesota; Dr. Wei Zhang, University of Minnesota; Mr. Karthik Bharathwaj for providing us with images that were used in this book.

We also thank our editor Sunil Nair and editorial assistant Sarah Gelson; project coordinator Laurie Schlags; project editor Amy Rodriguez at Taylor and Francis/CRC Press for helping us during the proofreading and publication process.

Introduction

This book is meant for upper level undergraduates, graduate students and researchers in various disciplines in science, technology and mathematics. The book covers both image acquisition and image processing. The knowledge of image acquisition will help readers to perform experiments more effectively and cost efficiently. The knowledge of image processing will help the reader to analyze and measure accurately. Through this book, the concepts of image processing will become ingrained using examples written using Python, long recognized as one of the easiest languages for non-programmers to learn.

Python is a good choice for teaching image processing because:

1. It is freely available and open source. Since it is a free software, all students will have access to it without any restriction.

2. It provides pre-packed installations available for all major platforms at no cost.

3. It is the high-level language of choice for scientists and engineers.

4. It is recognized as perhaps the easiest language to learn for non-programmers.

Due to new developments in imaging technology as well as the scientific need for higher resolution images, the image data sets are getting larger every year. Such large data sets can be analyzed quickly using a large number of computers. Closed source software like MATLAB® cannot be scaled to a large number of computers as the licensing cost is high. On the other hand, Python, being free and open-source software, can be scaled to thousands of computers at no cost. For these reasons, we strongly believe the future need for image processing for all students can be met using Python.

The book consists of three parts: Python programming, image processing, and image acquisition. Each of these parts consists of multiple chapters. The parts are self-contained. Hence, a user well versed in Python programming can skip Part I and read only Parts II and III. Each chapter contains many examples, detailed derivations, and working Python examples of the techniques discussed within. The chapters are also interspersed with practical tips on image acquisition and processing. The end of every chapter contains a summary of the important points discussed and a list of exercise problems to cement the reader's understanding.

Part I consists of introduction to Python, Python modules, reading and writing images using Python, and an introduction to images. Readers can skip or skim this part if they are already familiar with the material. This part is a refresher and readers will be directed to other resources as applicable.

In Part II, we discuss image processing and computer vision algorithms. The various chapters discuss pre- and post-processing using filters, affine transformation, segmentation, morphological operations, image measurements, neural network and convolutional neural network.

In Part III, we discuss image acquisition using various modalities such as X-ray, Computed Tomography (CT), Magnetic Resonance Imaging (MRI), light microscopy and electron microscopy. These modalities cover most of the common image acquisition methods used currently by researchers in academia and industry.

Details about exercises

The Python programming and image processing parts of the book contain exercises that test the reader's skills in Python programming, image processing, and integration of the two. Solutions to odd-numbered problems, example programs and images are available at https://github.com/zenr/IMAUP-book-ed-2.

Authors

Ravishankar Chityala, Ph.D. is currently working as Principal Engineer in Silicon Valley. He has more than eighteen years of experience in image processing. He teaches Python programming and Deep learning using Tensorflow at the University of California Santa Cruz, Silicon Valley Campus. Previously, he worked as an image processing consultant at the Minnesota Supercomputing Institute of the University of Minnesota. As an image processing consultant, Dr. Chityala had worked with faculty, students and staff from various departments in the scientific, engineering and medical fields at the University of Minnesota, and his interaction with students had made him aware of their need for greater understanding of and ability to work with image processing and acquisition. Dr. Chityala co-authored "Essential Python" (Essential Education, California, 2018), also contributed to the writing of *Handbook of Physics in Medicine and Biology* (CRC Press, Boca Raton, 2009, Robert Splinter). His research interests are in image processing, machine learning and deep learning.

Sridevi Pudipeddi, Ph.D. has eleven years experience teaching undergraduate courses. She teaches Machine Learning with Python and Python for Data Analysis at the University of California Berkeley at San Francisco campus. Dr. Pudipeddi's research interests are in machine learning, applied mathematics and image and text processing. Python's simple syntax and its vast image processing capabilities, along with the need to understand and quantify important experimental information through image acquisition, have inspired her to co-author this book. Dr. Pudipeddi co-authored "Essential Python" (Essential Education, California, 2018).

List of Symbols and Abbreviations

$\sum$	summation
θ	angle
$|x|$	absolute value of x
e	2.718281
$*$	convolution
$\log$	logarithm base 10
$\ominus$	morphological erosion
$\oplus$	morphological dilation
$\circ$	morphological open
$\bullet$	morphological close
$\cup$	union
λ	wavelength
E	energy
h	Planck's constant
c	speed of light
μ	attentuation coefficient
γ	gyromagnetic ratio
NA	numerical aperture
ν	frequency
dx	differential
∇	gradient
$\frac{\partial}{\partial x}$	derivative along x-axis
$\nabla^2 = \Delta$	Laplacian
$\int$	integration
CDF	cumulative distribution function
CT	computed tomography
DICOM	digital imaging and communication in medicine
JPEG	joint photographic experts group
MRI	magnetic resonance imaging
PET	positron emission tomography
PNG	portable network graphics
PSF	point spread function
RGB	red, green, blue channels
TIFF	tagged image file format

Part I

Introduction to Images and Computing using Python

Chapter 1

Introduction to Python

1.1 Introduction

Before we begin discussion on image acquisition and processing using Python, we will provide an overview of the various aspects of Python. This chapter focuses on some of the basic materials covered by many other books [Bea09], [Het10], [Lut06], [Vai09] and also from the book, "Essential Python" ([PC18]), a book from the authors of this book. If you are already familiar with Python and are currently using it, then you can skip this chapter.

We begin with an introduction to Python. We will then discuss the installation of Python with all the modules using the Anaconda distribution. Once the installation has been completed, we can begin exploring the various features of Python. We will quickly review the various data structures such as list, dictionary, and tuples and statements such as for-loop, if-else, iterators and list comprehension.

1.2 What Is Python?

Python is a popular high-level programming language. It can handle various programming tasks such as numerical computation, web development, database programming, network programming, parallel processing, etc.

Python is popular for various reasons including:

1. It is free.

2. It is available on all the popular operating systems such as Windows, Mac or Linux.

3. It is an interpreted language. Hence, programmers can test portions of code on the command line before incorporating it into their program. There is no need for compiling or linking.

4. It gives the ability to program faster.

5. It is syntactically simpler than C/C++/Fortran. Hence it is highly readable and easier to debug.

6. It comes with various modules that are standard or can be installed in an existing Python installation. These modules can perform various tasks like reading and writing various files, scientific computation, visualization of data, etc.

7. Programs written in Python can be run on various OS or platforms with little or no change.

8. It is a dynamically typed language. Hence the data type of variables does not have to be declared prior to their use, making it easier for people with less coding experience.

9. It has a dedicated developer and user community and is kept up to date.

Although Python has many advantages that have made it one of the most popular interpreted languages, it has a couple of drawbacks that are discussed below:

1. Since its focus is on the ability to program faster, the speed of execution suffers. A Python program might be 10 times or more slower (say) than an equivalent C program, but it will contain

fewer lines of code and can be programmed to handle multiple data types easily. This drawback in the Python code can be overcome by converting the computationally intensive portions of the code to C/C++ or by the appropriate use of data structure and modules.

2. Indentation of the code is not optional. This makes the code readable. However, a code with multiple loops and other constructs will be indented to the right, making it difficult to read the code.

1.3 Python Environments

There are several Python environments from which to choose. Some operating systems like Mac, Linux, Unix, etc. have a built-in interpreter. The interpreter may contain all modules but is not turn-key ready for scientific computing. Specialized distributions have been created and sold to the scientific community, pre-built with various Python scientific modules. When using these distributions, the users do not have to individually install scientific modules. If a particular module that is of interest is not available in the distribution, it can be installed. One of the most popular distributions is Anaconda [Ana20b]. The instructions for installing Anaconda distribution can be found at `https://www.anaconda.com/distribution/`.

1.3.1 Python Interpreter

The Python interpreter built into most operating systems can be started by simply typing `python` in the terminal window. When the interpreter is started, a command prompt ($>>>$) appears. Python commands can be entered at the prompt for processing. For example, in Mac, when the built-in Python interpreter is started, an output similar to the one shown below appears:

```
(base) mac:ipaup ravi$ python
Python 3.7.3 | packaged by conda-forge |
(default, Dec  6 2019, 08:36:57)
[Clang 9.0.0 (tags/RELEASE_900/final)] :: Anaconda, Inc.
on darwin
Type "help", "copyright", "credits" or "license"
for more information.
>>>
```

Notice that in the example above, the Python interpreter is version 3.7.3. It is possible that you might have a different version.

1.3.2   Anaconda Python Distribution

The Anaconda Python Distribution [Ana20a] provides programmers with close to 100 of the most popular scientific Python modules like scientific computation, linear algebra, symbolic computing, image processing, signal processing, visualization, integration of C/C++ programs to Python etc. It is distributed and maintained by Continuum Analytics. It is available for free for academics and is available for a price to all others. In addition to the various modules built into Anaconda, programmers can install other modules using the conda [Ana20b] package manager, without affecting the main distribution.

To access Python from the command line, start the 'Anaconda Prompt' executable and then type **python**.

1.4   Running a Python Program

Using any Python interpreter (built-in or from a distribution), you can run your program using the command at the operating system (OS) command prompt. If the file **firstprog.py** is a Python file that needs to be executed, then type the following command on the OS command prompt.

```
>> python firstprog.py
```

The >> is the terminal prompt and >>> represents the Python prompt.

The best approach to running Python programs under any operating systems is to use an Integrated Development Environment like IDLE or Spyder as it provides an ability to edit the file and also run it under the same interface.

1.5 Basic Python Statements and Data Types

Indentation

In Python, a code block is indicated by indentation. For example in the code below, we first print a message, 'We are computing squares of numbers between 0 and 9'. Then we loop through values in the range of 0 to 9 and store it in the variable 'i' and also print the square of 'i'. Finally we print the message, 'We completed the task ...' at the end.

In other languages, the code block under the for-loop would be identified with a pair of curly braces {}. However, in Python we do not use curly braces. The code block is identified by moving the line print(i*i) four spaces to the right. You can also choose to use a tab instead.

```
print('Computing squares of numbers between 0 and 9')
for i in range(10):
    print(i*i)
print('Completed the task ...')
```

There is a significant disadvantage to indentation especially to new Python programmers. A code containing multiple for-loops and if-statements will be indented farther to the right making the code unreadable. This problem can be alleviated by reducing the number of for-loops and if-statements. This not only makes the code readable but also reduces computational time. This can be achieved by programming using data structures like lists, dictionary, and sets appropriately.

Comments

Comments are an important part of any programming language. In Python, a single line comment is denoted by a hash # at the beginning of a line. Multiple lines can be commented by using triple quoted strings (triple single quotes or triple double quotes) at the beginning and at the end of the block.

```
# This is a single line comment

'''
This is
a multiline
comment
'''


# Comments are a good way to explain the code.
```

Variables

Python is a dynamic language and hence you do not need to specify the variable type as in C/C++. Variables can be imagined as containers of values. The values can be an integer, float, string, lists, tuples, dictionary, set, etc.

```
>>> a = 1
>>> a = 10.0
>>> a = 'hello'
```

In the above example the integer value of 1, float value of 10.0, and a string value of `hello` for all cases are stored in the same variable. However, only the last assigned value is the current value for a.

Operators

Python supports all the common arithmetic operators such as $+, -, *, /$. It also supports the common comparison operators such as $>, <, ==, ! =, >=, <=$, etc. In addition, through various modules Python provides many operators for performing trigonometric, mathematical, geometric operations, etc.

Loops

The most common looping construct in Python is the `for-loop` statement, which allows iterating through the collection of objects. Here is an example:

```
>>> for i in range(1,5):
...         print(i)
```

In the above example the output of the `for-loop` are the numbers from 1 to 5. The range function allows us to create values starting from 1 and ending with 5. Such a concept is similar to the `for-loop` normally found in C/C++ or most programming languages.

The real power of `for-loop` lies in its ability to iterate through other Python objects such as lists, dictionaries, sets, strings, etc. We will discuss these Python objects in more detail subsequently.

```
>>> a = ['python', 'scipy']
>>> for i in a:
...         print(i)
```

In the program above, the `for-loop` iterates through each element of the list and prints it.

In the next program, the content of a dictionary is printed using the `for-loop`. A dictionary with two keys `lang` and `ver` is defined. Then, using the `for-loop`, the various keys are iterated and the corresponding values are printed.

```
>>> a = {
          'lang':'python'
          'ver': '3.6.6'
      }
>>> for keys in a:
...     print(a[key])
```

The discussion about using a `for-loop` for iterating through the various lines in a text file, such as comma separated value file, is postponed to a later section.

if-else statement

The `if-else` is a popular conditional statement in all programming languages including Python. An example of `if-elif-else` statement is shown below.

```
if a<10:
    print('a is less than 10')
elif a<20:
    print('a is between 10 and 20')
else:
    print('a is greater than 20')
```

The `if-else` statement conditionals do not necessarily have to use the conditional operators such as $<, >, ==$, etc.

For example, the following `if` statement is legal in Python. This `if` statement checks for the condition that the list d is not empty.

```
>>> d = [ ]
>>> if d:
...     print('d is not empty')
... else:
...     print('d is empty')

d is empty
```

In the above code, since d is empty, the else clause is true and we enter the else block and print d is empty.

1.5.1 Data Structures

The real power of Python lies in the liberal usage of its data structure. The common criticism of Python is that it is slow compared to C/C++. This is especially true if multiple for-loops are used in programming Python. This can be alleviated by appropriate use of data structures such as lists, tuples, dictionary and sets. We describe each of these data structures in this section.

Lists

Lists are similar to arrays in C/C++. But, unlike arrays in C/C++, lists in Python can hold objects of any type such as int, float, string and including another list. Lists are mutable, as their size can be changed by adding or removing elements. The following examples will help show the power and flexibility of lists.

```
>>> a = ['python','scipy', 3.6]
>>> a.pop(-1)
3.6
>>> print(a)
a = ['python','scipy']
>>> a.append('numpy')
>>> print(a)
['python','scipy', 3.6]
>>> print(a[0])
python
>>> print(a[-1])
numpy
>>> print(a[0:2])
['python','scipy']
```

In the first line, a new list is created. This list contains two strings and one floating-point number. In the second line, we use the pop function to remove the last element (index = −1). The popped element is printed to the terminal. After the pop, the list contains only two elements instead of the original three. We use append, and insert a new element, "numpy" to the end of the list. Finally, in the next two commands we print the value of the list in index 0 and the last position indicated by using "−1" as the index. In the last command, we introduce slicing and obtain a new list that contains only the first two values of the list. This indicates that one can operate on the list using methods such as pop, insert, or remove and also using operators such as slicing.

A list may contain another list. Here is an example. We will consider the case of a list containing four numbers and arranged to look like a matrix.

```
>>> a = [[1,2],[3,4]]
>>> print(a[0])
[1,2]
>>> print(a[1])
[3,4]
>>> print(a[0][0])
1
```

In line 1, we define a list of the list. The values [1, 2] are in the first list and the values [3, 4] are in the second list. The two lists are combined to form a 2D list. In the second line, we print the value of the first element of the list. Note that this prints the first row or the first list and not just the first cell. In the fourth line, we print the value of the second row or the second list. To obtain the value of the first element in the first list, we need to index the list as given in line 6. As you can see, indexing the various elements of the list is as simple as calling the location of the element in the list.

Although the list elements can be individually operated, the power of Python is in its ability to operate on the entire list at once using list methods and list comprehensions.

List functions/methods

Let us consider the list that we created in the previous section. We can sort the list using the sort method as shown in line 2. The sort method does not return a list; instead, it modifies the current list. Hence the existing list will contain the elements in a sorted order. If a list contains both numbers and strings, Python sorts the numerical values first and then sorts the strings in alphabetical order.

```
>>> a = ['python','scipy','numpy']
>>> a.sort()
>>> a
['numpy','python','scipy']
```

List comprehensions

A list comprehension allows building a list from another list. Let us consider the case where we need to generate a list of squares of numbers from 0 to 9. We will begin by generating a list of numbers from 0 to 9. Then we will determine the square of each element.

```
>>> a = list(range(10))
>>> print(a)
[0, 1, 2, 3, 4, 5, 6, 7, 8, 9]
>>> b = [x*x for x in a]
[0, 1, 4, 9, 16, 25, 36, 49, 64, 81]
>>> b = []
>>> for x in a:
        b.append(x*x)
>>> print(b)
[0, 1, 4, 9, 16, 25, 36, 49, 64, 81]
```

In line 1, a list is created containing values from 0 to 9 using the function "range" and the print command is given in line 2. In line number 4, list comprehension is performed by taking each element in 'a' and multiplying by itself. The result of the list comprehension is shown in line 5. The same operation can be performed by using lines 6 to 8 but the list comprehension approach is compact in syntax as it eliminates two lines of code, one level of indentation, and a for-loop. It is also faster when applied to a large list.

For a new Python programmer, the list comprehension might seem daunting. The best way to understand and read a list comprehension is by imagining that you will first operate on the for-loop and then begin reading/writing the left part of the list comprehension. In addition to applying for-loop using list comprehension, you can also apply logical operations like if-else.

Tuples

Tuples are similar to lists except that they are not mutable, i.e., the length and the content of the tuple cannot be changed at runtime. Syntactically, the list uses [ ] while tuples use ( ). Similar to lists, tuples may contain any data type including other tuples. Here are a few examples:

```
>>> a = (1,2,3,4)
>>> print(a)
(1,2,3,4)
>>> b = (3,)
>>> c = ((1,2),(3,4))
```

In line 1, we define a tuple containing four elements. In line 4, we define a tuple containing only one element. Although the tuple contains only one element, we need to add the trailing comma, so that Python understands it as a tuple. Failure to add a comma at the end of this tuple will result in the value 3 being treated as an integer and not a tuple. In line 5, we create a tuple inside another tuple.

Sets

A set is an unordered collection of unique objects. To create a set, we need to use the function set or the operator {}. Here are some examples:

```
>>> s1 = set([1,2,3,4])
>>> s2 = set((1,1,3,4))
>>> print(s2)
set([1,3,4])
```

In line 1, we create a set from a list containing four values. In line 2, we create a set containing a tuple. The elements of a set need to be unique. Hence when the content of s2 is printed, we notice that the duplicates have been eliminated. Sets in Python can be operated using many of the common mathematical operations on sets such as union, intersection, set difference, symmetric difference, etc.

Since sets do not store repeating values and since we can convert lists and tuples to sets easily, they can be used to perform useful operations faster which otherwise would involve multiple loops and conditional statements. For example, a list containing only unique values can be obtained by converting the list to a set and back to a list. Here is an example:

```
>>> a = [1,2,3,4,3,5]
>>> b = set(a)
>>> print(b)
set([1,2,3,4,5])
>>> c = list(b)
>>> print(c)
[1,2,3,4,5]
```

In line 1, we create a list containing six values with one duplicate. We convert the list into a set by using the set() function. During this process, the duplicate value 3 is eliminated. We can then convert the set back to list using the list() function.

Dictionaries

Dictionaries store key-value pairs. A dictionary is created by enclosing a key-value pair inside { }.

```
>>> a = {
                'lang':'python'
                'ver': '3.6.6'
        }
```

Any member of the dictionary can be accessed using [ ] operator.

```
>>> print a['lang']
python
```

To add a new key,

```
>>> a['creator'] = 'Guido Von Rossum'
>>> print(a)
{'lang': 'python', 'ver': '3.6.6', 'creator': 'Guido Von
Rossum'}
```

In the example above, we added a new key called creator and stored the string, "Guido Von Rossum."

In certain instances, the dictionary membership needs to be tested using the 'in' operator. To obtain a list of all the dictionary keys, use the keys() method.

1.5.2   File Handling

This book is on image processing; however, it is important to understand and be able to include in your code, reading and writing text files so that the results of computation can be written or the input parameters can be read from external sources. Python provides the ability to read and write files. It also has functions, methods and modules for reading specialized formats such as the comma separated value (csv) file, Microsoft Excel (xls) format, etc. We will look into each method in this section.

The following code reads a csv file as a text file.

```
>>> fo = open('myfile.csv')
>>> for i in fo.readlines():
...   print(i)
>>> fo.close()
Python,3.6.6

Django, 3.0.5

Apache, 2.4
```

The first line opens a file and returns a new file object which is stored in the file object "fo." The method **readlines** in line 2, reads all the lines of input. The for-loop then iterates over each of those lines, and prints. The file is finally closed using the close method.

The output of the print command is a string. Hence, string manipulation using methods like split, strip, etc., needs to be applied in order to extract elements of each column. Also, note that there is an extra newline character at the end of each print statement.

Reading CSV files

Instead of reading a csv file as a text file, we can use the csv module.

```
>>> import csv
>>> for i in csv.reader(open('myfile.csv')):
...     print(i)
['Python', '3.6.6']
['Django', ' 3.0.5']
['Apache', ' 2.4']
```

The first line imports the csv module. The second line opens and reads the content of the csv file using the reader function in the csv module. In every iteration of the loop, the content of one line of the csv file is returned and stored in the variable 'i'. Finally, the value of 'i' is printed.

Reading Excel files

Microsoft Excel files can be read and written using the `openpyxl` module. The openpyxl module has to be installed before we can use it. To install this module, you can go to the Python prompt and type 'pip install openpyxl'. Alternately, the module can be installed by selecting the Environment tab in Anaconda Navigator. Here is a simple example of reading an Excel file using the `openpyxl` module.

```
from openpyxl import load_workbook
wb = load_workbook('myfile.xlsx')
for sheet in wb:
    for row in sheet.values:
        for col in row:
            print(col, end=' | ')
        print()
```

In line 2, the `open_workbook()` function is used to read the file. We loop through all sheets in the file. In this particular example, there is only one sheet. Then we loop through each row in the sheet followed by looping through every column. In the print function where the value in a column is printed, we are indicating that a column must be separated from the next column by a | (pipe symbol). Finally the last print function adds a new line, so that the next row can be printed in a new line.

```
Date | Time |
2020-01-02 00:00:00 | 10:15:00 |
2020-01-05 00:00:00 | 11:00:00 |
2020-01-07 00:00:00 | 15:00:00 |
```

1.5.3   User-Defined Functions

A function is a reusable section of code that may take input and may or may return an output. If there is any block of code that will be called many times, it is advisable to convert it to a function. Calls can then be made to this function.

A Python function can be created using the def keyword. Here is an example:

```python
import math
def circleproperties(r):
    area = math.pi*r*r;
    circumference = 2*math.pi*r
    return area, circumference


a, c = circleproperties(5) # Radius of the circle is 5
print("Area and Circumference of the circle are", a, c)
```

The function circleproperties takes in one input argument, the radius (r). The return statement at the end of the function definition passes the computed values (in this case, area and circumference) to the calling function. To invoke the function, use the name of the function and provide the radius value as an argument enclosed in parentheses. Finally, the area and circumference of the circle are displayed using the print function call.

The variables area and circumference have local scope. Hence the variables cannot be invoked outside the body of the function. It is possible to pass on variables to a function that have global scope using the global statement.

When we run the above program, we get the following output:

```
Area and Circumference of the circle are 78.539 31.415
```

1.6 Summary

- Python is a popular high-level programming language. It is used for most common programming tasks such as scientific computation, text processing, building dynamic websites, etc.

- Python distributions such as Anaconda Python distribution are pre-built with many scientific modules and enable scientists to focus on their research instead of installation of modules.

- Python, like other programming languages, uses common relational and mathematical operators, comment statements, for-loops, if-else statements, etc.

- To program like a Pythonista, use lists, sets, dictionary and tuples liberally.

- Python can read most of the common text formats like CSV, Microsoft Excel, etc.

1.7 Exercises

1. If you are familiar with any other programming language, list the differences between that language and Python.

2. Write a Python program that will print numbers from 10 to 20 using a for-loop.

3. Create a list of state names such as states = ['Minnesota', 'Texas', 'New York', 'Utah', 'Hawaii']. Add another entry 'California' to the end of the list. Then, print all the values of this list.

4. Print the content of the list from Question 3 and also the corresponding index using the list's **enumerate** method in the `for-loop`.

5. Create a 2D list of size 3-by-3 with the following elements: 1, 2, 3|4, 5, 6|6, 7, 8

6. It is easy to convert a list to a set and vice versa. For example, a list '*mylist* = [1, 1, 2, 3, 4, 4, 5]' can be converted to a set

using the command `newset = set(mylist)`. The set can be converted back to a list using `newlist = list(newset)`. Compare the contents of mylist and newlist. What do you infer?

7. Look up documentation for the `join` method and join the contents of the list ['Minneapolis','MN','USA'] and obtain the string 'Minneapolis, MN, USA.'

8. Consider the following Python code:

```
a = [1,2,3,4,2,3,5]
b = []
for i in a:
    if i>2:
        b.append(i)
print(b)
```

Rewrite the above code using list comprehension and reduce the number of lines.

Chapter 2

Computing using Python Modules

2.1 Introduction

We discussed the basics of Python in the previous chapter. We learned that Python comes with various built-in batteries or modules. These batteries or modules perform various specialized operations. The modules can be used to perform computation, database management, web server functions etc. Since this book is focused on creating scientific applications, we limit our focus to Python modules that allow computation such as the scipy, numpy, matplotlib, Python Imaging Library (PIL), and scikit packages. We discuss the relevance of each of these modules and explain their use with examples. We also discuss creation of new Python modules.

2.2 Python Modules

A number of scientific Python modules have been created and are available in the Python distributions used in this book. Some of the most popular modules relevant to this book's scope are:

1. **numpy**: A powerful library for manipulating arrays and matrices.

2. **scipy**: Provides functions for performing higher-order mathematical operations such as filtering, statistical analysis, image processing, etc.

3. **matplotlib**: Provides functions for plotting and other forms of visualization.

4. **Python Imaging Library**: Provides functions for basic image reading, writing and processing.

5. **scikits**: An add-on package for scipy. The modules in scikit are meant to be added to scipy after development.

In the following sections, we will describe these modules in detail. Please refer to [BS13],[Bre12],[Idr12] to learn more.

2.2.1   Creating Modules

A module is a Python file containing multiple functions or classes and other optional components. All these functions and classes share a common namespace, namely, the name of the module file. For example, the following program is a valid Python module.

```
# filename: examplemodules.py
version = '1.0'

def printpi():
    print('The value of pi is 3.1415')
```

A function named 'printpi' and a variable called 'version' was created in this module. The function performs the simple operation of printing the value of π.

2.2.2   Loading Modules

To load this module, use the following command in the Python command line or in a Python program. The word "examplemodules" is the name of the module file.

```
>>> import examplemodules
```

Once the module is loaded, the function can be run using the command below. The first command prints the value of pi along with a label while the second command prints the version number.

```
>>> examplemodules.printpi()
The value of pi is 3.1415
```

```
>>> examplemodules.version
'1.0'
```

The example module shown above has only one function. A module may contain multiple functions or classes.

In the first example, the datetime module is loaded. In the example however, we are only interested in obtaining the current date using date.today().

```
>>> import datetime
>>> print(datetime.date.today())
2020-02-08
```

In the second example, only the necessary function (date) in the datetime module that is needed is loaded. For large modules, it is recommended to import only the necessary functionality to make the code readable.

```
>>> from datetime import date
>>> print(date.today())
2020-02-08
```

In the third example, we import all the functions in a given module using *. Once imported, the file name (in this case "date") that contains the function (in this case "today()") needs to be specified. This import method is typically not recommended, as it can result in namespace collision. For example, it is ambiguous if the date functionality is in the datetime module or if it is from some other import statement.

```
>>> from datetime import *
>>> print(date.today())
2020-02-08
```

In the fourth example, we import a module (in this case numpy) and rename it to something shorter such as np. This is known as aliasing. This will reduce the number of characters that need to be typed and consequently the lines of code to maintain.

```
>>> import numpy as np
>>> np.ones([3,3])
array([[ 1.,   1.,   1.],
       [ 1.,   1.,   1.],
       [ 1.,   1.,   1.]])
```

For the purpose of this book, we focus on only a few modules that are detailed below.

2.3   Numpy

A numpy module adds the ability to manipulate arrays and matrices using a library of mathematical functions. Numpy is derived from the now defunct modules Numeric and Numarray. Numeric was the first attempt to provide the ability to manipulate arrays but it was very slow for computation on large arrays. Numarray, on the other hand, was too slow on small arrays. The code base was combined to create numpy.

Numpy has functions and routines to perform linear algebra, random sampling, polynomials, financial functions, set operations, etc. Since this book is focused on image processing and since images are arrays, we will be using the matrix manipulation capabilities of numpy. The second module that we will be discussing is scipy, which internally uses numpy for its matrix manipulation.

The drawback of Python compared to C or $C++$ is the speed of execution. This is in part due to its interpreted execution. A Python program for numeric computation written with a similar construct to a C program using a for-loop will perform considerably poorly. The best method of programming Python for faster execution is to use numpy and scipy modules. The following program illustrates the problem in programming using a for-loop. In this program, the value of π is calculated using the Gregory-Leibniz-Madhava method. The method can be expressed as

$$\pi = 4 * \left\{ 1 - \frac{1}{3} + \frac{1}{5} - \frac{1}{7} + \frac{1}{9} \cdots \right\}. \tag{2.1}$$

The corresponding program is shown below. In the program, we perform the following operations:

1. Create the numerator and denominator separately using numpy's linspace and ones functions. The details of the two functions can be found in the numpy documentation.

2. Begin a while-loop and find the ratio between the elements of numerator and denominator and the corresponding sum.

3. Multiply the value of the sum by 4 to obtain the value of π.

4. Print the time for completing the operation.

```
import time
import numpy as np

def main():
    noofterms = 10000000
    # Calculate the denominator.
    # First few terms are 1,3,5,7 ...
# den is short for denominator
    den = np.linspace(1,noofterms*2,noofterms)
```

```python
# Calculate the numerator
# The first few terms are
# 1, -1, 1, -1 ...
# num is short for numerator
num = np.ones(noofterms)
for i in range(1,noofterms):
    num[i] = pow(-1,i)

counter = 0
sum_value = 0

t1 = time.process_time()
while counter<noofterms:
    sum_value += (num[counter]/den[counter])
    counter = counter + 1
pi_value = sum_value*4.0
print("pi_value is: %f" % pi_value)
t2 = time.process_time()
# Determine the time for computation
timetaken = t2-t1
print("Time taken is: %f seconds" % timetaken)

if __name__ == '__main__':
    main()
```

When we run the above program, we get the following output:
pi_value is 3.141593
Time taken is 6.203125 seconds

The program below is the same as the one above except for step 3, where instead of calculating the sum of the ratio of the numerator and denominator using a while-loop or for-loop, we calculate using numpy's sum function.

```python
import numpy as np
import time

def main():
    # No of terms in the series
    noofterms = 1000000
    # Calculate the denominator.
    # First few terms are 1,3,5,7 ...
    # den is short for denominator
    den  = np.linspace(1,noofterms*2,noofterms)
    # Calculate the numerator.
    # The first few terms are 1, -1, 1, -1 ...
    # num is short for numerator
    num = np.ones(noofterms)
    for i in range(1,noofterms):
        num[i] = pow(-1,i)
    # Find the ratio and sum all the fractions
    # to obtain pi value
    # Start the clock
    t1 = time.process_time()
    pi_value =  sum(num/den)*4.0
    print("pi_value is: %f" % pi_value)
    t2 = time.process_time()
    # Determine the time for computation
    timetaken = t2-t1
    print("Time taken is: %f seconds" % timetaken)

if __name__ == '__main__':
    main()
```

The output of the program with numpy's sum is:

pi_value is 3.141592

Time taken is 0.328125 seconds

The first program took 6.203125 seconds while the second program took 0.328125 seconds, for an approximate speed-up of 18. Although this example performs a fairly simple computation, a real-world problem that takes a few weeks to solve can be completed in a few days with the appropriate use of numpy and scipy. Also, the program is elegant without the indentation used in the while-loop.

2.3.1   Numpy Array or Matrices?

Numpy manipulates mathematical matrices and vectors and hence computes faster than a traditional for-loop that manipulates scalars. In numpy, there are two types of mathematical matrix classes: arrays and matrices. The two classes have been designed for similar purposes but arrays are more general-purpose and n-dimensional, while matrices facilitate faster linear algebra calculations. Some of the differences between arrays and matrices are listed below:

1. Matrix objects have rank 2, while arrays have rank > 2.

2. Matrix objects can be multiplied by using the * operator, while the same operator on an array performs element-by-element multiplication. The *dot()* needs to be used for performing multiplication on arrays.

3. Array is the default datatype on numpy.

The arrays are used more often in numpy and other modules that use numpy for their computation. The matrix and array can be interchanged but it is recommended to use arrays.

2.4 Scipy

Scipy is a library of functions, programs and mathematical tools for scientific programming in Python. It uses numpy for its internal computation. Scipy is an extensive library that allows programming of different mathematical applications such as integration, optimization, Fourier transforms, signal processing, statistics, multi-dimensional image processing etc.

Travis Oliphant, Eric Jones and Pearu Peterson merged their modules to form scipy in 2001. Since then, many volunteers all over the world have participated in maintaining scipy.

As stated in Section 2.2, loading modules can be expensive both in CPU and memory used. This is especially true for large packages like scipy that contain many submodules. In such cases, load only the specific submodule.

```
>>> from scipy import ndimage
>>> import scipy.ndimage as im
```

In the first command, only the ndimage submodule is loaded. In the second command, the ndimage module is loaded as im.

The subsequent chapters will use scipy for all their image processing computations and hence details will be discussed later.

2.5 Matplotlib

Matplotlib is a 2D/3D plotting library for Python. It is designed to use the numpy datatype. It can be used for generating plots inside a Python program. An example demonstrating the features of matplotlib is shown in Figure 2.1.

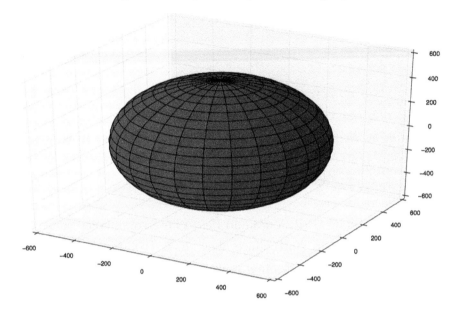

FIGURE 2.1: Example of a plot generated using matplotlib.

2.6   Python Imaging Library

Python Imaging Library (PIL) is a module for reading, writing and processing image files. It supports most of the common image formats like JPEG, PNG, TIFF, etc. In a subsequent section, PIL will be used for reading and writing images.

2.7   Scikits

Scikits is a short form for scipy toolkits. It is an additional package that can be used along with scipy tools. An algorithm is programmed in scikits if:

1. The algorithm is still under development and is not ready for prime time in scipy.

2. The package has a license that is not compatible with scipy.

3. Scipy is a general-purpose scientific package in Python. Thus, it is designed so that it is applicable to a wide array of fields. If a package is deemed specialized for a certain field, it continues to be part of scikits.

Scikits consists of modules from various fields such as environmental science, statistical analysis, image processing, microwave engineering, audio processing, boundary value problem, curve fitting, quantum computing, etc.

In this book, we will focus only on the image processing routines in scikits named scikit-image. The scikit-image routine contains algorithms for input/output, morphology, object detection and analysis, etc.

```
>>> from skimage import filters
>>> import skimage.filters as fi
```

In the first command, only the filters submodule is loaded. In the second command, the filters module is loaded as fi.

2.8   Python OpenCV Module

The Open Source Computer Vision Library (OpenCV) [Ope20a] is an image processing, computer vision and machine learning software library. It has more than 2000 algorithms for processing image data. It has a large user base and is used extensively in academic institutions, commercial organizations, and government agencies. It provides binding for common programming languages such as C, C++, Python, etc. Python binding is used in a few examples in this book.

To import the Python OpenCV module, type the following in the command line:

```
>>> import cv2
```

2.9   Summary

- Various Python modules for performing image processing were discussed. They are numpy, scipy, matplotlib, Python Imaging Library, Python OpenCV, and scikits.

- The module has to be loaded before using functions that are specific to that module.

- In addition to using existing Python modules, user-defined modules can be created.

- Numpy modules add the ability to manipulate arrays and matrices using a library of high-level mathematical functions. Numpy has two data structures for storing mathematical matrices. They are arrays and matrices. An array is more versatile than a matrix, and is more commonly used in numpy and also in all the modules that use numpy for computation.

- Scipy is a library of programs and mathematical tools for scientific programming in Python.

- Scikits is used for the development of new algorithms that can later be incorporated into scipy.

2.10   Exercises

1. Python is an open-source and free software. Hence, there are many modules created for image processing. Perform research and discuss some of the benefits of each module over another.

2. Although this book is on image processing, it is important to combine the image processing operation with other mathematical

operations such as optimization, statistics, etc. Perform research about combining image processing with other mathematical operations.

3. Why is it more convenient to arrange the various functions as modules?

4. You are provided a CSV file containing a list of full path to file names of various images. The file has only one column with multiple rows. Each row contains the path to one file. You need to read the file name and then read the image as well. The method for reading a CSV file was shown in Chapter 1.

5. Modify the program from Question 4 to read a Microsoft Excel file instead.

6. Create a numpy array of size 5-by-5 containing all random values. Determine the transpose and inverse of this matrix.

Chapter 3

Image and Its Properties

3.1   Introduction

We begin this chapter with an introduction to images, image types, and data structures in Python. Image processing operations can be imagined as a workflow similar to Figure 3.1. The workflow begins with reading an image. The image is then processed using either low-level or high-level operations. Low-level operations operate on individual pixels. Such operations include filtering, morphology, thresholding, etc. High-level operations include image understanding, pattern recognition, etc. Once processed, the image(s) are either written to disk or visualized. The visualization may be performed during the course of processing as well. We will discuss this workflow and the functions using Python as an example.

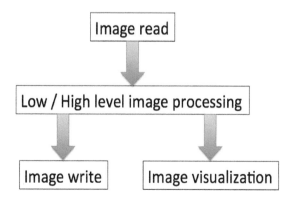

FIGURE 3.1: Image processing work flow.

3.2   Image and Its Properties

In the field of medical imaging, the images may span all spatial dimensions (x-, y- and x-axis) and also the time dimension. Hence it is common to find images in 3D, and in some cases such as cardiac CT, images in 4D. In the case of optical microscopy, the images of the same specimen may be acquired at various emission and excitation wavelengths. Such images will span multiple channels and may have more than 4 dimensions. We begin the discussion by clarifying some of the mathematical terms that are used in this book.

For simplicity, let us assume the images that will be discussed in this book are 3D volumes. A 3D volume (I) can be represented mathematically as

$$\alpha = I \longrightarrow \mathbb{R} \text{ and } I \subset \mathbb{R}$$

Thus, every pixel in the image has a real number as its value. However, in reality as it is easier to store integers than to store floats; most images have integers for pixel values.

3.2.1   Bit-Depth

The pixel range of a given image format is determined by its bit-depth. The range is $[0, 2^{bitdepth} - 1]$. For example, an 8-bit image will have a range of $[0, 2^8 - 1] = [0, 255]$. An image with higher bit-depth needs more storage in disk and memory. Most of the common photographic formats such as JPEG, PNG, etc. use 8 bits for storage and only have positive values.

Medical and microscope images use a higher bit-depth, as scientific applications demand higher accuracy. A 16-bit medical image will have values in the range $[0, 65535]$ for a total number $65536 \; (= 2^{16})$ values. For a 16-bit image that has both positive and negative pixel values, the range is $[-32768, +32767]$. The total number of values in this case is

65536 ($= 2^{16}$) or a bit-depth of 16. A good example of such an image is a CT DICOM image.

Scientific image formats store the pixel values at high precision, not only for accuracy, but also to ensure that physical phenomena records are not lost. In CT, for example, a pixel value of > 1000 indicates bone. If the image is stored in 8-bit, the pixel value of bone would be truncated at 255 and hence the information will be permanently lost. In fact, the most significant pixels in CT have intensity > 255 and hence need higher bit-depth.

There are a few image formats that store images at even higher bit-depth such as 32 or 64. For example, a JPEG image containing RGB (3 channels) will have a bit-depth of 8 for each channel and hence has a total bit-depth of 24. Similarly, a TIFF microscope image with 5 channels (say) with each channel at 16-bit depth will have a total bit-depth of 80.

3.2.2 Pixel and Voxel

A pixel in an image can be thought of as a bucket that collects light or electrons depending on the type of detector used. A single pixel in an image spans a distance in the physical world. For example, in Figure 3.2, the arrows indicate the width and height of a pixel placed adjacent to three other pixels. In this case, the width and height of this pixel is 0.5 mm. Thus in a physical space, traversing a distance of 0.5 mm is equivalent to traversing 1 pixel in the pixel space. For all practical purposes, we can assume that detectors have square pixels, i.e., the pixel width and pixel height are the same.

The pixel size could be different for different imaging modalities and different detectors. For example, the pixel size is greater for CT compared to micro-CT.

In medical and microscope imaging, it is more common to acquire 3D images. In such cases, the pixel size will have a third dimension,

1 pixel width = 0.5mm

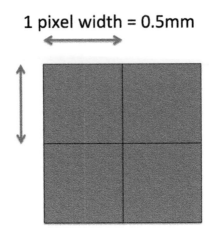

FIGURE 3.2: Width and height of pixel in physical space.

namely the pixel depth. The term pixel is generally applied to 2D and is replaced by voxel in 3D images.

Most of the common image formats like DICOM, nifti, and some microscope image formats contain the voxel size in their header. Hence, when such images are read in a visualization or image processing program, an accurate analysis and visualization can be performed. But if the image does not have the information in the header or if the visualization or image processing program cannot read the header properly, it is important to use the correct voxel size for analysis.

Figure 3.3 illustrates the problem of using an incorrect voxel size in visualization. The left image is the volume rendering of an optical coherence tomography image with incorrect voxel size in the z-direction. The right image is the volume rendering of the same image with correct voxel size. In the left image, it can be seen clearly that the object is highly elongated in the z-direction. In addition, the undulations at the top of the volume and the five hilly structures at the top are also made prominent by the incorrect voxel size. The right image has the same shape and size as the original object. The problem not only affects visualization but also any measurements performed on the volume.

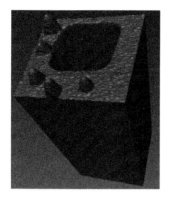

(a) Volume rendering with incorrect voxel size. The 3D is elongated in the z-direction.

(b) Volume rendering with correct voxel size.

FIGURE 3.3: An example of volume rendering with correct and incorrect voxel size.

3.2.3 Image Histogram

A histogram is a graphical depiction of the distribution of pixel value in an image. The image in Figure 3.4 is a histogram of an image. The x-axis is the pixel value and the y-axis is the frequency or the number of pixels with the given pixel value. In the case of an integer-based image such as JPEG, whose values span $[0, 255]$, the number of values in the x-axis will be 256. Each of these 256 values is referred to as a "bin." Several bins can also be used in the x-axis. In the case of images containing floating-point values, the bins will have a range of values.

Histograms are a useful tool in determining the quality of the image. A few observations can be made in Figure 3.4:

1. The left side of the histogram corresponds to lower pixel values. Hence if the frequency at lower pixel values is very high, it indicates that some of the pixels might be missing from that end,

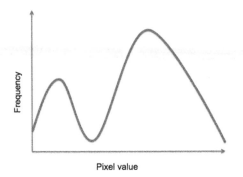

FIGURE 3.4: An example of a histogram.

i.e., there are values farther left of the first pixel that were not recorded in the image.

2. An ideal histogram should have close to zero frequency for the lower pixel values.

3. The right side of the histogram corresponds to higher pixel values. Hence, if the frequency at higher pixel values is very high, it indicates saturation, i.e., there might be some pixels to the right of the highest value that were never recorded.

4. An ideal histogram should have close to zero frequency for the higher pixel values.

5. The above histogram is bi-modal. The trough between the two peaks is the pixel value that can be used for segmentation by thresholding. But not all images have bi-modal histograms; hence there are many techniques for segmentation using histograms. We will discuss some of these techniques in Chapter 8, "Segmentation."

3.2.4   Window and Level

The human eye can view a large range of intensity values, while modern displays are severely limited in their capabilities.

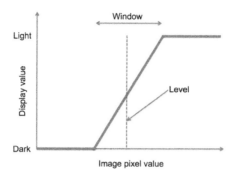

FIGURE 3.5: Window and level.

Image viewing applications display the pixel value after a suitable transformation due to the fact that displays have a lower intensity range than the intensity range in an image. One example of the transformation, namely window-level, is shown in Figure 3.5. Although the computer selects a transformation, the user can modify it by changing the window range and the level. The window allows modifying the contrast of the display while the level changes the brightness of the display.

3.2.5   Connectivity: 4 or 8 Pixels

The usefulness of this section will be more apparent with the discussion of convolution in Chapter 7, "Fourier Transform." During the convolution operation, a mask or kernel is placed on top of an image pixel. The final value of the output image pixel is determined using a linear combination of the value in the mask and the pixel value in the image. The linear combination can be calculated for either 4-connected pixels or 8-connected pixels. In the case of 4-connected pixels shown in Figure 3.6, the process is performed on the top, bottom, left and right pixels. In the case of 8-connected pixels, the process is performed in addition on the top-left, top-right, bottom-left and bottom-right pixels.

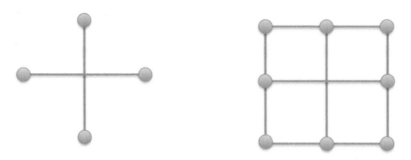

FIGURE 3.6: An example of 4 and 8 pixel connectivity.

3.3   Image Types

There are more than 100 image formats. Some of these formats, such as JPEG, GIF, PNG, etc., are used for photographic images. Formats such as DICOM, NIFTI, and Analyze AVW are used in medical imaging. Formats such as TIFF, ICS, IMS, etc., are used in microscope imaging. In the following sections, we discuss some of these formats.

3.3.1   JPEG

JPEG stands for the Joint Photographic Experts Group, a joint committee formed to add images to text terminals. Its extension is .jpg or .jpeg. It is one of the most popular formats due to its ability to compress the data significantly with minimal visual loss. In the initial days of the World Wide Web, JPEG became popular as it helped save bandwidth in image data transfer. It is a lossy format, that compresses data using Discrete Cosine Transform (DCT). The parameters of compression can be tuned to minimize the loss in detail. Since JPEG stores image data after transforming them using DCT, it is not very suitable for storing images that contain fine structures such as lines, curves, etc. Such images are better stored as PNG or TIFF. The JPEG images can be viewed using viewers that are built into most computers. Since JPEG images can be compressed, image standards such as TIFF and DICOM may use JPEG compression to store the image data when compression is needed.

3.3.2 TIFF

TIFF stands for Tagged Image File Format. Its extension is .tif or .tiff. The latest version of the TIFF standards is 6.0. It was created in the 80's for storing and encoding scanned documents. It was developed by Aldus Corporation, which was later acquired by Adobe Systems. Hence, the copyright for TIFF standards is held by Adobe Systems.

Originally it was developed for single-bit data but today's standards allow storage of 16-bit and even floating-point data. Charged Couple Device (CCD) cameras used in scientific experiments acquire images at more than 12-bit resolution and hence TIFF images that store high precision are used extensively. The TIFF images can be stored internally using JPEG lossy compression or can be stored with lossless compression such as LZW.

It is popular in the microscope community for the fact that it has higher bit-depth (> 12 bits) per pixel per channel and also for its ability to store a sequence of images in a single TIFF file. The latter is sometimes referred as 3D TIFF. Most of the popular image processing software for the microscope community can read most forms of TIFF images. Simple TIFF images can be viewed using viewers built into most computers. The TIFF images generated from scientific experiments are best viewed using applications that are specialized in that domain.

3.3.3 DICOM

Digital Imaging and Communication in Medicine (DICOM) is a standard format for encoding and transmitting medical CT and MRI data. This format stores the image information along with other data like patient details, acquisition parameters etc. DICOM images are used by doctors in various disciplines such as radiology, neurology, surgery, cardiology, oncology, etc. There are more than 20 DICOM committees that meet and update the standards 4 or 5 times a year. It is managed by the National Electrical Manufacturers Association (NEMA), which owns the copyright of the DICOM standards.

DICOM format uses tested tools such as JPEG, MPEG, TCP/IP for its internal working. This allows easier deployment and creation of DICOM tools. DICOM standards also define the transfer of images, storage, and other allied workflows. Since DICOM standards have become popular, many image processing readers and viewers have been created to read, process and write images.

DICOM images have header as well as image data similar to other image formats. But, unlike other format headers, the DICOM header contains not only information about the size of the image, pixel size, etc., but also patient information, physician information, imaging parameters, etc. The image data may be compressed using various techniques like JPEG, lossless JPEG, run length encoding (RLE), etc. Unlike other formats, DICOM standards define both the data format and also the protocol for transfer.

The listing below is a partial example of a DICOM header. The patient and doctor information have been either removed or altered for privacy. Section 0010 contains patient information, section 0009 details the CT machine used for acquiring the image, and section 0018 details the parameter of acquisition, etc.

```
0008,0022   Acquisition Date: 20120325
0008,0023   Image Date: 20120325
0008,0030   Study Time: 130046
0008,0031   Series Time: 130046
0008,0032   Acquisition Time: 130105
0008,0033   Image Time: 130108
0008,0050   Accession Number:
0008,0060   Modality: CT
0008,0070   Manufacturer: GE MEDICAL SYSTEMS
0008,0080   Institution Name: ----------------------
```

```
0008,0090   Referring Physician's Name: XXXXXXX
0008,1010   Station Name: CT01_OC0
0008,1030   Study Description: TEMP BONE/ST NECK W
0008,103E   Series Description: SCOUTS
0008,1060   Name of Physician(s) Reading Study:
0008,1070   Operator's Name: ABCDEF
0008,1090   Manufacturer's Model Name: LightSpeed16
0009,0010   ---: GEMS_IDEN_01
0009,1001   ---: CT_LIGHTSPEED
0009,1002   ---: CT01
0009,1004   ---: LightSpeed16
0010,0010   Patient's Name: XYXYXYXYXYXYX
0010,0020   Patient ID: 213831
0010,0030   Patient's Birth Date: 19650224
0010,0040   Patient's Sex: F
0010,1010   Patient's Age:
0010,21B0   Additional Patient History:
            ? MASS RIGHT EUSTACHIAN TUBE
0018,0022   Scan Options: SCOUT MODE
0018,0050   Slice Thickness: 270.181824
0018,0060   kVp: 120
0018,0090   Data Collection Diameter: 500.000000
0018,1020   Software Versions(s): LightSpeedverrel
0018,1030   Protocol Name: 3.2 SOFT TISSUE NECK
0018,1100   Reconstruction Diameter:
0018,1110   Distance Source to Detector: 949.075012
0018,1111   Distance Source to Patient: 541.000000
0018,1120   Gantry/Detector Tilt: 0.000000
0018,1130   Table Height: 157.153000
0018,1140   Rotation Direction: CW
0018,1150   Exposure Time: 2772
0018,1151   X-ray Tube Current: 10
0018,1152   Exposure: 27
```

```
0018,1160  Filter Type: BODY FILTER
0018,1170  Generator Power: 1200
0018,1190  Focal Spot(s): 0.700000
0018,1210  Convolution Kernel: STANDARD
```

The various software that can be used to manipulate DICOM images can be found online. We will classify these software based on the user requirements.

The user might need:

1. A simple viewer with limited manipulation like ezDICOM [Ror20].

2. A viewer with ability to manipulate images and perform rendering like Osirix [SAR20].

3. A viewer with image manipulation capability and also extensible with plugins like ImageJ.

1. **ezDICOM:** This is a viewer that provides sufficient functionality that allows users to view and save DICOM files without installing any other complex software in their system. It is available only for Windows OS. It can read DICOM files and save them in other file formats. It can also convert image files to analyze format.

2. **Osirix:** This is a viewer with extensive functionality and is available free, but unfortunately it is available only in MacOSX. Like other DICOM viewers, it can read and store files in different file formats and as movies. It can perform Multi-Planar Reconstruction (MPR), 3D surface rendering, 3D volume rendering, and endoscopy. It can also view 4D DICOM data. The surface rendered data can also be stored as VRML, STL files, etc.

3. **ImageJ:** ImageJ was funded by the National Institutes of Health (NIH) and is available as open source. It is written in Java and users can add their own Java classes or plugins. It is available

in all major operating system like Windows, Linux, UNIX, Mac, etc. It can read all DICOM formats and can store the data in various common file formats and also as movies. The plugins allow various image processing operations. Since the plugins can be easily added, the complexity of the image processing operation is limited only by the user's knowledge of Java. Since ImageJ is a popular image processing software, a brief introduction is presented in Appendix C, "Introduction to ImageJ."

3.4 Data Structures for Image Analysis

Image data is generally stored as a mathematical matrix. So in general, a 2D image of size 1024-by-1024 is stored in a matrix of the same size. Similarly, a 3D image is stored in a 3D matrix. In numpy, a mathematical matrix is called a numpy array. As we will be discussing in the subsequent chapters, the images are read and stored as a numpy array and then processed using either functions in a Python module or user-defined functions.

Since Python is a dynamically typed language (i.e., no defining data type), it will determine the data type and size of the image at run time and store appropriately.

3.5 Reading, Writing and Displaying Images

3.5.1 Reading Images

After a lot of research, we decided to use Python's computer vision module, OpenCV [Ope20a] for reading and writing images, the PIL module's Image for reading images, and Matplotlib's pyplot to display images.

OpenCV is imported as cv2. We use imread function to read an image, which returns an ndarray. The cv2.imread supports the following file formats:

- Windows bitmaps: bmp, dib

- JPEG files: jpeg, jpg, jpe

- JPEG 2000 files: jp2

- Portable Network Graphics: png

- Portable image format: pbm, pgm, ppm

- TIFF files: tiff, tif

Below is the cv2's code snippet for reading images.

```
import cv2

# Reading image and converting it into an ndarray
img = cv2.imread('Picture1.png')

# Converting img to grayscale
img_grayscale = cv2.cvtColor(img, cv2.COLOR_BGR2GRAY)
```

We import cv2 and use the cv2.imread function to read the image as an ndarray. To convert a color image to grayscale, we use the function cvtColor whose first argument is the ndarray of an image and the second argument is cv2.COLOR_BGR2GRAY. This argument, cv2.COLOR_BGR2GRAY, converts an RGB image, which is a three-channel ndarray to grayscale, which is a single-channel ndarray using the following formula:

$$y = 0.299 * R + 0.587 * G + 0.114 * B \qquad (3.1)$$

Another way to read images would be using PIL module's Image class. Below is the code snippet for reading images using PIL's Image.

```
from PIL import Image
import numpy as np

# Reading image and converting it into grayscale.
img = Image.open('Picture2.png').convert('L')
# convert PIL Image object to numpy array
img = np.array(img)

# Performing image processing on img.
img2 = image_processing(img)

# Converting ndarray to image for saving using PIL.
im3 = Image.fromarray(img2)
```

In the above code, we import Image from the PIL module. We open the 'Picture.png' image and convert a three-channel image to a single-channel grayscale by using convert('L') and the result is a PIL Image object. We then convert this PIL Image object to a numpy ndarray using the np.array function because most image processing modules in Python can only handle a numpy array and not a PIL Image object. After performing some image processing operation on this ndarray, we convert the ndarray back to an image using Image.fromarray, so that it can be saved or visualized.

3.5.2 Reading DICOM Images using pyDICOM

We will use pyDICOM [Mas20], a module in Python to read or write or manipulate DICOM images. The process for reading DICOM images is similar to JPEG, PNG, etc. Instead of using cv2, the pyDICOM module is used. The pyDICOM module is not installed by default in the distributions. Please refer to the pyDICOM documentation at [Mas20]. To read a DICOM file, the DICOM module is first imported. The file is then read using the "read_file" function.

```
import dicom
ds = dicom.read_file("ct_abdomen.dcm")
```

3.5.3   Writing Images

Throughout this book, to write or save images we will use cv2.imwrite. The cv2.imwrite function supports the following file formats:

- JPEG files: *.jpeg, *.jpg, *.jpe

- Portable Network Graphics: *.png

- Portable image format: *.pbm, *.pgm, *.ppm

- TIFF files: *.tiff, *.tif

Here is an example code snippet where we read an image and write an image. The imwrite function takes the file name and the ndarray of an image as input. The file format is identified using the file extension in the file name.

```
import cv2
img = cv2.imread('image1.png')


# cv2.imwrite will take an ndarray.
cv2.write('file_name', img)
```

In the subsequent chapters, we will continue to use the above approach for writing or saving images.

3.5.4   Writing DICOM Images using pyDICOM

To write a DICOM file, the DICOM module is first imported. The file is then written using the "write_file" function. The input to the function is the name of the DICOM file and also the array that needs to be stored.

```
import dicom
datatowrite = ...
dicom.write_file("ct_abdomen.dcm",datatowrite)
```

3.5.5 Displaying Images

Throughout this book, to display images, we will use Matplotlib.pyplot. Below is a sample code snippet that reads and displays an image.

```
import cv2
import matplotlib.pyplot as plt


# cv2.imread will read the image and convert it into an
ndarray.
img = cv2.imread('image1.png')


# We import matplotlib.pyplot to display an image in
grayscale.
# If gray is not supplied the image will be displayed
in color.
plt.imshow(img, 'gray')
plt.show()
```

We are importing cv2 and matplotlib.pyplot modules. Notice that we are aliasing matplotlib.pyplot as plt. We use cv2.imread to read an image and we use plt.imshow to display the image. As we want a grayscale image to be displayed, we provide a string 'gray' to the plt.imshow function.

Note: We can also display a DICOM image using plt.imshow because pyDICOM's read_file also returns a data object that contains the image data as an ndarray.

3.6 Programming Paradigm

As described in the introductory section, the workflow (Figure 3.1) for image processing begins with reading an image and finally ends with

either writing the image to file or visualizing it. The image processing
operations are performed between the reading and writing or visualizing
the image. In this section, the code snippet that will be used for reading
and writing or visualizing the image is presented. This code snippet will
be used in most of the programs presented in this book.

Below is a sample code where cv2 and matplotlib are used.

```
# cv2 module's imread to read an image as an ndarray.
# cv2 module's imwrite to write an image.
import cv2
import matplotlib.pyplot as plt


img = cv2.imread('image1.png')


# Converting img to grayscale (if needed).
img_grayscale = cv2.cvtColor(img, cv2.COLOR_BGR2GRAY)


# We process img_grayscale and obtain img_processed.
# The function image_processing can perform any image
# processing or computer vision operation
img_processed = image_processing(img_grayscale)


# cv2.imwrite will take an ndarray and store it.
cv2.write('file_name.png', img_processed)


# We import matplotlib.pyplot to display an image in
grayscale.
plt.imshow(img_processed, 'gray')
plt.show()
```

In the above code, the cv2 module is imported. Then matplotlib
.pyplot is imported as plt. We use cv2.imread to read image1.png
and return an ndarray. We use cv2.cvtColor along with the argument
cv2.COLOR_BGR2GRAY to convert img, which is a three-channel
ndarray to a single-channel ndarray and we store it in img_grayscale.

We perform some image processing (assuming that the function image_processing already exists) on img_grayscale and we assign it to img_processed. We save img_processed using cv2.write, which converts the ndarray to an image. To visualize, we use plt.imshow. The plt.imshow function takes the ndarray as a necessary input argument and image type as an optional argument. In this case, for image type, we chose gray.

Below is another sample code where PIL and matplotlib are used instead of cv2 for reading and writing images.

```
# PIL module to read and save an image.
from PIL import Image
import matplotlib.pyplot as plt


# Opening image and converting it into grayscale.
img = Image.open('image2.png').convert('L')
# convert PIL Image object to numpy array
img = np.array(img)


# We process img_grayscale and obtain img_processed
img_processed = image_processing(img)


# Converting ndarray to a PIL Image.
img_out = Image.fromarray(img_processed)


# Save the image to a file.
img_out.save('file_name.png')


# Display the image in grayscale
plt.imshow(img_processed, 'gray')
plt.show()
```

In the above code, Image class is imported from PIL. Then matplotlib is imported as plt. We use Image.open to read the image and in the same line, using convert('L'), the image is converted from a three-channel image to a single-channel Image object. We use the np.array function to convert the PIL image to ndarray. We then perform some image processing (assuming that the function image_processing already exists) on img and we assign it to img_processed. We use Image.fromarrray to convert img_processed, which is an ndarray, to a PIL Image object. We save img_processed using the save method in the PIL Image class. To visualize, we use plt.imshow. The plt.imshow function takes the ndarray as a necessary input argument and image type as an optional argument. In this case, for image type, we chose gray.

3.7　Summary

- Image processing is preceded by reading an image file. It is then followed by either writing the image to file or visualization.

- An image is stored generally in the form of matrices. In Python, it is processed as a numpy n-dimensional array or ndarray.

- An image has various properties like bit-depth, pixel/voxel size, histogram, window-level, etc. These properties affect the visualization and processing of images.

- There are hundreds of image formats created to serve the needs of the image processing community. Some of these formats like JPEG, PNG, etc. are used generally for photographs while DICOM, Analyze AVW, and NIFTI are used for medical image processing.

- In addition to processing these images, it is important to view these images using graphical tools such as ezDicom, Osirix, ImageJ, etc.

- Reading and writing images can be performed using many methods. One such method was presented in this chapter. We will continue to use this method in all the subsequent chapters.

3.8   Exercises

1. An image of size 100-by-100 has isotropic pixel size of 2-by-2 microns. The number of pixels in the foreground is 1000. What is the area of the foreground and background in $microns^2$?

2. A series of images are used to create a volume of data. There are 100 images each of size 100-by-100. The voxel size is 2-by-2-by-2 microns. Determine the volume of the foreground in $microns^3$ given the number of pixels in the foreground is 10,000.

3. A histogram plots the frequency of occurrence of the various pixel values. This plot can be converted to a probability density function or pdf, so that the y-axis is the probability of the various pixel values. How can this be accomplished?

4. To visualize window or level, open an image in any image processing software (such as ImageJ). Adjust window and level. Comment on the details that can be seen for different values of window and level.

5. There are specialized formats for microscope images. Conduct research on these formats.

Part II

Image Processing using Python

Chapter 4

Spatial Filters

4.1 Introduction

So far we have covered the basics of Python and its scientific modules. In this chapter, we begin our journey of learning image processing. The first concept we will master is filtering, which is at the heart of image quality and further processing.

We associate filters (such as a water filter) with removing undesirable impurities. Similarly, in image processing, a filter removes undesirable impurities which includes noise. In some cases, the impurities might be visually distracting and in some cases might produce error in further processing. Some filters are also used to suppress certain features in an image and highlight others. For example, the first derivative and second derivative filters that we will discuss are used to determine or enhance edges in an image.

There are two types of filters: linear filters and non-linear filters. Linear filters include mean, Laplacian and Laplacian of Gaussian. Non-linear filters include median, maximum, minimum, Sobel, Prewitt and Canny filters.

Image enhancement can be accomplished in two domains: spatial and frequency. The spatial domain constitutes all the pixels in an image. Distances in the image (in pixels) correspond to real distances in micrometers, inches, etc. The domain over which the Fourier transformation of an image ranges is known as the frequency domain of the image. We begin with image enhancement techniques in the spatial domain. Later in Chapter 7, "Fourier Transform," we will discuss image enhancement using frequency or Fourier domain.

The Python modules that are used in this chapter are scikits and scipy. Scipy documentation can be found at [Sci20c], scikits documentation can be found at [Sci20a], and scipy ndimage documentation can be found at [Sci20d].

4.2   Filtering

As a water filter removes impurities, an image processing filter removes undesired features (such as noise) from an image. Each filter has a specific utility and is designed to either remove a type of noise or to enhance certain aspects of the image. We will discuss many filters along with their purposes and their effects on images.

For filtering, a filter or mask is used. It is usually a two-dimensional square window that moves across the image affecting only one pixel at a time. Each number in the filter is known as a coefficient. The coefficients in the filter determine the effects of the filter and consequently the output image. Let us consider a 3-by-3 filter, F, given in Table 4.1.

TABLE 4.1: A 3-by-3 filter.

F_1	F_2	F_3
F_4	F_5	F_6
F_7	F_8	F_9

If (i, j) is the pixel in the image, then a sub-image around (i, j) of the same dimension as the filter is considered for filtering. The center of the filter is placed to overlap with (i, j). The pixels in the sub-image are multiplied with the corresponding coefficients in the filter. This yields a matrix of the same size as the filter. The matrix is simplified using a mathematical equation to obtain a single value that will replace the pixel value in (i, j) of the image. The exact nature of the mathematical equation depends on the type of filter. For example, in the case of a mean filter, the value of $F_i = \frac{1}{N}$, where N is the number of elements in the filter. The filtered image is obtained by repeating the process of placing the filter on every pixel in the image, obtaining the single value and replacing the pixel value in the original image. This process of sliding a filter window over an image is called convolution in the spatial domain.

Let us consider the following sub-image from the image, I, centered at (i, j)

TABLE 4.2: A 3-by-3 sub-image.

$I(i-1, j-1)$	$I(i-1, j)$	$I(i-1, j+1)$
$I(i, j-1)$	$I(i, j)$	$I(i, j+1)$
$I(i+1, j-1)$	$I(i+1, j)$	$I(i+1, j+1)$

The convolution of the filter given in Table 4.1 with the sub-image in Table 4.2 is given as follows:

$$
\begin{aligned}
I_{new}(i, j) = {} & F_1 * I(i-1, j-1) + F_2 * I(i-1, j) + F_3 * I(i-1, j+1) \\
& + F_4 * I(i, j-1) + F_5 * I(i, j) + F_6 * I(i, j+1) \\
& + F_7 * I(i+1, j-1) + F_8 * I(i+1, j) + F_9 * I(i+1, j+1)
\end{aligned}
$$
$$(4.1)$$

where $I_{new}(i, j)$ is the output value at location (i, j). This process has to be repeated for every pixel in the image. Since the filter plays an important role in the convolution process, the filter is also known as the convolution kernel.

The convolution operation has to be performed at every pixel in the image including pixels at the boundary of the image. When the filter is placed on the boundary pixels, a portion of the filter will lie outside the boundary. Since the pixel values do not exist outside the boundary, new values have to be created prior to convolution. This process of creating pixel values outside the boundary is called padding. The padded pixels can be assumed to be either zero or a constant value. Other padding options such as nearest neighbor or reflect create padded pixels using pixel values in the image. In the case of zeros, the padded pixels are all zeros. In the case of constant, the padded pixels take a specific value. In the case of reflect, the padded pixels take the value of the last row/s or column/s. The padded pixels are considered only for convolution and will be discarded after convolution.

Let us consider an example to show different padding options. Figure 4.1(a) is a 7-by-7 input image to be convolved using a 3-by-5 filter with the center of the filter at $(1, 2)$. In order to include boundary pixels for convolution, we pad the image with one row above and one row below and two columns to the left and two columns to the right. In general the size of the filter dictates the number of rows and columns that will be padded to the image.

- **Zero padding**: All padded pixels are assigned a value of zero (Figure 4.1(b)).

- **Constant padding**: A constant value of 5 is used for all padded pixels (Figure 4.1(c)). The constant value can be chosen based on the type of image that is being processed.

- **Nearest neighbor**: The values from the last row or column (Figure 4.1(d)) are used for padding.

- **Reflect**: The values from the last row or column (Figure 4.1(e)) are reflected across the boundary of the image.

- **Wrap**: In the wrap option given in Figure 4.1(f), the first row (or column) after the boundary takes the same values as the first row (or column) in the image and so on.

4.2.1   Mean Filter

In mathematics, functions are classified into two groups, linear and non-linear. A function f is said to be linear if

0	2	5	7	3	10	9
11	1	4	6	8	2	0
0	12	10	9	7	4	5
1	9	7	8	13	11	0
5	10	14	6	2	1	1
7	6	11	3	13	8	4
3	9	6	12	7	10	5

(a) A 7-by-7 input image.

0	0	0	0	0	0	0	0	0	0	0
0	0	0	2	5	7	3	10	9	0	0
0	0	11	1	4	6	8	2	0	0	0
0	0	0	12	10	9	7	4	5	0	0
0	0	1	9	7	8	13	11	0	0	0
0	0	5	10	14	6	2	1	1	0	0
0	0	7	6	11	3	13	8	4	0	0
0	0	3	9	6	12	7	10	5	0	0
0	0	0	0	0	0	0	0	0	0	0

(b) Padding with zeros.

5	5	5	5	5	5	5	5	5	5	5
5	5	0	2	5	7	3	10	9	5	5
5	5	11	1	4	6	8	2	0	5	5
5	5	0	12	10	9	7	4	5	5	5
5	5	1	9	7	8	13	11	0	5	5
5	5	5	10	14	6	2	1	1	5	5
5	5	7	6	11	3	13	8	4	5	5
5	5	3	9	6	12	7	10	5	5	5
5	5	5	5	5	5	5	5	5	5	5

(c) Padding with a constant.

0	0	0	2	5	7	3	10	9	9	10
0	0	0	2	5	7	3	10	9	9	9
11	11	11	1	4	6	8	2	0	0	0
0	0	0	12	10	9	7	4	5	5	5
1	1	1	9	7	8	13	11	0	0	0
5	5	5	10	14	6	2	1	1	1	1
7	7	7	6	11	3	13	8	4	4	4
3	3	3	9	6	12	7	10	5	5	5
3	3	3	9	6	12	7	10	5	5	5

(d) Padding with nearest neighbor.

2	0	0	2	5	7	3	10	9	9	10
2	0	0	2	5	7	3	10	9	9	10
1	11	11	1	4	6	8	2	0	0	2
12	0	0	12	10	9	7	4	5	5	4
9	1	1	9	7	8	13	11	0	0	11
10	5	5	10	14	6	2	1	1	1	1
6	7	7	6	11	3	13	8	4	4	8
9	3	3	9	6	12	7	10	5	5	10
9	3	3	9	6	12	7	10	5	5	10

(e) Padding with reflect option.

5	10	3	9	6	12	7	10	5	3	9
9	10	0	2	5	7	3	10	9	0	2
0	2	11	1	4	6	8	2	0	11	1
5	4	0	12	10	9	7	4	5	0	12
0	11	1	9	7	8	13	11	0	1	9
1	1	5	10	14	6	2	1	1	5	10
4	8	7	6	11	3	13	8	4	7	6
5	10	3	9	6	12	7	10	5	3	9
9	10	0	2	5	7	3	10	9	0	2

(f) Padding with wrap option.

FIGURE 4.1: An example of different padding options.

$$f(x + y) = f(x) + f(y) \tag{4.2}$$

Otherwise, f is non-linear. A linear filter is an extension of the linear function.

An excellent example of a linear filter is the mean filter. The coefficients of mean filter F (Table 4.1) are 1's. To avoid scaling the pixel intensity after filtering, the whole image is then divided by the number of pixels in the filter; in the case of a 3-by-3 sub-image we divide it by 9.

Unlike other filters discussed in this chapter, the mean filter does not have a scipy.ndimage module function. However, we can use the convolve function to achieve the intended result. The following is the signature of the Python function for convolve:

```
scipy.ndimage.filters.convolve(input, weights)
```

```
Necessary arguments:
  input is a numpy ndarray.

  weights is an ndarray consisting of
  coefficients of 1s for the mean filter.

Optional arguments:
  mode determines the method for handling the array
  border by padding. Different options are: constant,
  reflect, nearest, mirror, wrap. See above explanation.

  cval is a scalar value specified when the mode option
  is constant. The default value is 0.0.

  origin is a scalar that determines filter origin.
  The default value 0 corresponds to a filter
```

whose origin (reference pixel) is at the center.
In a 2D case, origin = 0 would mean (0,0).

Returns: output is an ndarray

The program explaining the usage of the mean filter is given below.
The filter (k) is an ndarray array of size 5-by-5 with all values =
1/25. The filter is then convolved using the "convolve" function from
scipy.ndimage.filters.

```
import cv2
import numpy as np
import scipy.ndimage

# Opening the image using cv2.
a = cv2.imread('../Figures/ultrasound_muscle.png')
# Converting the image to grayscale.
a = cv2.cvtColor(a, cv2.COLOR_BGR2GRAY)

# Initializing the filter of size 5 by 5.
# The filter is divided by 25 for normalization.
k = np.ones((5,5))/25
# performing convolution
b = scipy.ndimage.filters.convolve(a, k)
# Writing b to a file.
cv2.imwrite('../Figures/mean_output.png', b)
```


Figure 4.2(a) is an ultrasound image of muscle. Notice that the
image contains noise. The mean filter of size 5-by-5 is applied to remove
the noise. The output is shown in Figure 4.2(b). The mean filter effectively removed the noise but in the process blurred the image.

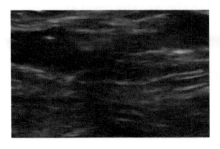
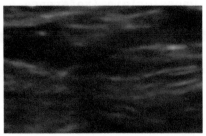

(a) Input image for mean filter.   (b) Output generated with a mean filter size (5,5).

FIGURE 4.2: Example of mean filter.

Advantages of the mean filter

- Removes noise.

- Enhances the overall quality of the image, i.e., mean filter brightens an image.

Disadvantages of the mean filter

- In the process of smoothing, the edges get blurred.

- Reduces the spatial resolution of the image.

If the coefficients of the mean filter are not all 1s, then the filter is a weighted mean filter. In the weighted mean filter, the filter coefficients are multiplied with the sub-image as in the non-weighted filter. After application of the filter, the image should be divided by the total weight for normalization.

4.2.2 Median Filter

Functions that do not satisfy Equation 4.2 are non-linear. The median filter is one of the most popular non-linear filters. A sliding window is chosen and is placed on the image at the pixel position (i, j).

All pixel values under the filter are collected. The median of these values is computed and is assigned to (i, j) in the filtered image. For example, consider a 3-by-3 sub-image with values 5, 7, 6, 10, 13, 15, 14, 19, 23. To compute the median, the values are arranged in ascending order, so the new list is: 5, 6, 7, 10, 13, 14, 15, 19, and 23. Median is a value that divides the list into two equal halves; in this case it is 13. So the pixel (i, j) will be assigned 13 in the filtered image. The median filter is most commonly used in removing salt-and-pepper noise and impulse noise. Salt-and-pepper noise is characterized by black and white spots randomly distributed in an image.

The following is the Python function for the median filter:

```
scipy.ndimage.filters.median_filter(input, size=None,
    footprint=None, mode='reflect', cval=0.0, origin=0)


Necessary arguments:
 input is the input image as an ndarray.

Optional arguments:
 size can be a scalar or a tuple. For example, if the
 image is 2D, size = 5 implies a 5-by-5 filter is
 considered. Alternately, the size can also be specified
 as size=(5,5).

 footprint is a boolean array of the same dimension as
 the size unless specified otherwise. The pixels in the
 input image corresponding to the points to the
 footprint with true values are considered for
 filtering.

 mode determines the method for handling the array
 border by padding. Options are: constant,
 reflect, nearest, mirror, wrap.
```

Returns: output image as an ndarray.

The Python code for the median filter is given below:

```
import cv2
import scipy.ndimage

# Opening the image and converting it to grayscale.
a = cv2.imread('../Figures/ct_saltandpepper.png')
# Converting the image to grayscale.
a = cv2.cvtColor(a, cv2.COLOR_BGR2GRAY)
# Performing the median filter.
b = scipy.ndimage.filters.median_filter(a, size=5)
# Saving b as median_output.png in Figures folder
cv2.imwrite('../Figures/median_output.png', b)
```

In the above code, $size = 5$ represents a filter (mask) of size 5-by-5. The image in Figure 4.3(a) is a CT slice of the abdomen with salt-and-pepper noise. The image is read using 'cv2.imread' and the ndarray returned is passed to the median_filter function. The output of the median_filter function is stored as a 'png' file. The output image is shown in Figure 4.3(b). The median filter efficiently removed the salt-and-pepper noise.

4.2.3   Max Filter

This filter enhances the bright points in an image. In this filter the maximum value in the sub-image replaces the value at (i, j). The Python function for the maximum filter has the same arguments as the median filter discussed above. The Python code for the max filter is given below.

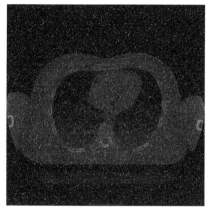 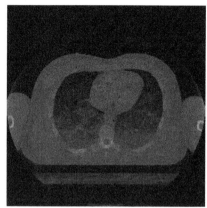

(a) Input image for median filter.   (b) Output generated with a filter size=(5,5).

FIGURE 4.3: Example of a median filter.

```
import scipy.misc
import scipy.ndimage
from scipy.misc.pilutil import Image

# opening the image and converting it to grayscale
a = Image.open('../Figures/wave.png').convert('L')
# performing maximum filter
b = scipy.ndimage.filters.maximum_filter(a, size=5)
# b is converted from an ndarray to an image
b = scipy.misc.toimage(b)
b.save('../Figures/maxo.png')
```

The image in Figure 4.4(a) is the input image for the max filter. The input image has a thin black boundary on the left, right and bottom. After application of the max filter, the white pixels have grown and hence the thin edges in the input image are replaced by white pixels in the output image as shown in Figure 4.4(b).

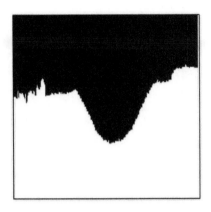
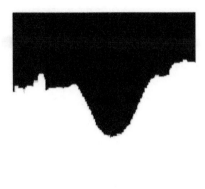

(a) Input image for max filter.   (b) Output image of max filter.

FIGURE 4.4: Example of max filter.

4.2.4 Min Filter

This filter is used to enhance the darkest points in an image. In this filter, the minimum value of the sub-image replaces the value at (i, j). The Python function for the minimum filter has the same arguments as the median filter discussed above. The Python code for the min filter is given below.

```
import cv2
import scipy.ndimage

# opening the image and converting it to grayscale
a = cv2.imread('../Figures/wave.png')

# performing minimum filter
b = scipy.ndimage.filters.minimum_filter(a, size=5)
# saving b as mino.png
cv2.imwrite('../Figures/mino.png', b)
```

After application of the min filter to the input image in Figure 4.5(a), the black pixels have grown and hence the thin edges in the input image are thicker in the output image as shown in Figure 4.5(b).

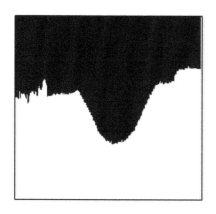 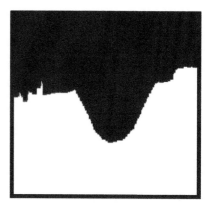

(a) Input image for min filter.   (b) Output image of min filter.

FIGURE 4.5: Example of min filter.

4.3 Edge Detection using Derivatives

Edges are a set of points in an image where there is a change of intensity between one side of that point and the other. From calculus, we know that the changes in intensity can be measured by using the first or second derivative. First, let us learn how changes in intensities affect the first and second derivatives by considering a simple image and its corresponding profile. This method will form the basis for using first and second derivative filters for edge detection. Interested readers can also refer to [MH80],[Mar72],[PK91] and [Rob77].

Figure 4.6(a) is the input image in grayscale. The left side of the image is dark while the right side is light. While traversing from left to right, at the junction between the two regions, the pixel intensity changes from dark to light. Figure 4.6(b) is the intensity profile across a horizontal cross-section of the input image. Notice that at the point of transition from dark region to light region, there is a change in intensity in the profile. Otherwise, the intensity is constant in the dark and light regions. For clarity, only the region around the point of transition is shown in the intensity profile (Figure 4.6(b)), first derivative (Figure 4.6(c)), and second derivative (Figure 4.6(d)) profiles. In the transition region, since the intensity profile is increasing, the first derivative is positive, while being zero in the dark and light regions. The first derivative has a maximum or peak at the edge. Since the first derivative is increasing before the edge, the second derivative is positive before the edge. Likewise, since the first derivative is decreasing after the edge, the second derivative is negative after the edge. Also, the second derivative is zero in dark and light regions as the corresponding first derivative is zero. At the edge, the second derivative is zero. This phenomenon of the second derivative changing the sign from positive before the edge to negative after the edge or vice versa is known as zero-crossing, as it takes a value of zero at the edge. The input image was simulated on a computer and does not have any noise. However, acquired images will have noise that may affect the detection of zero-crossing. Also, if the intensity changes rapidly in the profile, spurious edges will be detected by the zero-crossing. To prevent the issues due to noise or rapidly changing intensity, the image is pre-processed before application of a second derivative filter.

4.3.1   First Derivative Filters

An image is not a continuous function and hence derivatives are calculated using discrete approximations and not using functions. For the purpose of learning, let us look at the definition of the gradient of a continuous function and then extend it to discrete cases. If $f(x, y)$ is a continuous function, then the gradient of f as a vector is given by

$$\nabla f = \begin{bmatrix} f_x \\ f_y \end{bmatrix} \qquad (4.3)$$

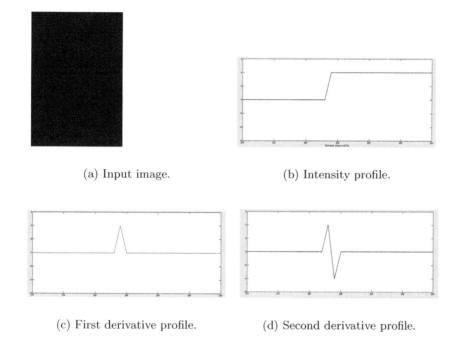

(a) Input image.

(b) Intensity profile.

(c) First derivative profile.

(d) Second derivative profile.

FIGURE 4.6: An example of zero-crossing.

where $f_x = \dfrac{\partial f}{\partial x}$ is known as the partial derivative of f with respect to x, (change of f along the horizontal direction) and $f_y = \dfrac{\partial f}{\partial y}$ is known as the partial derivative of f with respect to y, (change of f along the vertical direction). For more details refer to [Sch04]. The magnitude of the gradient is a scalar quantity and is given by

$$|\nabla f| = [(f_x)^2 + (f_y)^2]^{\frac{1}{2}} \qquad (4.4)$$

where $|z|$ is the norm of z.

For computational purposes, we will use the simplified version of the gradient given by Equation 4.5 and the angle given by Equation 4.6.

$$|\nabla f| = |f_x| + |f_y| \qquad (4.5)$$

$$\theta = \tan^{-1}\left(\frac{f_y}{f_x}\right) \qquad (4.6)$$

4.3.1.1   Sobel Filter

One of the most popular first derivative filters is the Sobel filter. The Sobel filter or mask is used to find horizontal and vertical edges as given in Table 4.3.

TABLE 4.3: Sobel masks for horizontal and vertical edges.

−1	−2	−1
0	0	0
1	2	1

−1	0	1
−2	0	2
−1	0	1

To understand how filtering is performed, let us consider a sub-image of size 3-by-3 given in Table 4.4 and multiply the sub-image with horizontal and vertical Sobel masks. The corresponding output is given in Table 4.5.

TABLE 4.4: A 3-by-3 subimage.

f_1	f_2	f_3
f_4	f_5	f_6
f_7	f_8	f_9

TABLE 4.5: Output after multiplying the sub-image with Sobel masks.

$-f_1$	$-2f_2$	$-f_3$
0	0	0
f_7	$2f_8$	f_9

$-f_1$	0	f_3
$-2f_4$	0	$2f_6$
$-f_7$	0	f_8

Since f_x is the partial derivative of f in the x direction, which is a change of f along the horizontal direction, the partial derivative can be obtained by taking the difference between the third row and the first row in the horizontal mask, so $f_x = (f_7 + 2f_8 + f_9) + (-f_1 - 2f_2 - f_3)$. Likewise, f_y is the partial derivative of f in the y direction, which is a change of f in the vertical direction; the partial derivative can be obtained by taking the difference between the third column and the first column in the vertical mask, so $f_y = (f_3 + 2f_6 + f_9) + (-f_1 - 2f_4 - f_7)$.

Using f_x and f_y f_5, the discrete gradient at f_5 (Equation 4.7) can be calculated.

$$|f_5| = |f_7 + 2f_8 + f_9 - f_1 - 2f_2 - f_3| + |f_3 + 2f_6 + f_9 - f_1 - 2f_4 - f_7| \quad (4.7)$$

The important features of the Sobel filter are:

- The sum of the coefficients in the mask image is 0. This means that the pixels with constant grayscale are not affected by the derivative filter.

- The side effect of derivative filters is creation of additional noise. Hence, coefficients of $+2$ and -2 are used in the mask image to produce smoothing.

The following is the Python function for the Sobel filter:

```
scipy.ndimage.sobel(image)

Necessary arguments:
  image is an ndarray with one or three channels.

Returns: output is an ndarrray.
```

The Python code for the Sobel filter is given below.

```python
import cv2
from scipy import ndimage

# Opening the image.
a = cv2.imread('../Figures/cir.png')
# Converting a to grayscale .
a = cv2.cvtColor(a, cv2.COLOR_BGR2GRAY)
```

```
# Performing Sobel filter.
b = ndimage.sobel(a)
# Saving b.
cv2.imwrite('../Figures/sobel_cir.png', b)
```

As can be seen in the code, the image, 'cir.png' is read using cv2.imread. The ndarray 'a' is then passed to the scipy.ndimage.sobel function to produce the Sobel edge enhanced image which is then written to file.

4.3.1.2 Prewitt Filter

Another popular first derivative filter is Prewitt [Pre70]. The masks for the Prewitt filter are given in Table 4.6.

TABLE 4.6: Prewitt masks for horizontal and vertical edges.

-1	-1	-1
0	0	0
1	1	1

-1	0	1
-1	0	1
-1	0	1

As in the case of the Sobel filter, the sum of the coefficients in Prewitt is also 0. Hence this filter does not affect pixels with constant grayscale. However, the filter does not reduce noise like the Sobel filter.

For Prewitt, the Python function's argument is similar to the Sobel function's argument.

Let us consider an example to illustrate the effect of filtering an image using both Sobel and Prewitt. The image in Figure 4.7(a) is a CT slice of a human skull near the nasal area. The output of the Sobel and Prewitt filters is given in Figures 4.7(b) and 4.7(c). Both filters have successfully created the edge image.

Slightly modified Sobel and Prewitt filters can be used to detect one or more types of edges. Sobel and Prewitt filters to detect diagonal edges are given in Tables 4.7 and 4.8.

To detect vertical and horizontal edges for Sobel and Prewitt filters we will use filters from the module skimage.

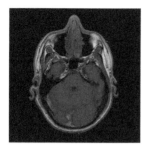

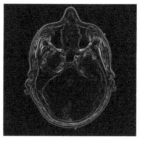

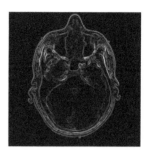

(a) A cross-section of human skull.

(b) Output of Sobel.

(c) Output of Prewitt.

FIGURE 4.7: Example for Sobel and Prewitt.

TABLE 4.7: Sobel masks for diagonal edges.

0	1	2
−1	0	1
−2	−1	0

−2	−1	0
−1	0	1
0	1	2

TABLE 4.8: Prewitt masks for diagonal edges.

0	1	1
−1	0	1
−1	−1	0

−1	−1	0
−1	0	1
0	1	1

- The function filters.sobel_v computes vertical edges using the Sobel filter.

- The function filters.sobel_h computes horizontal edges using the Sobel filter.

- The function filters.prewitt_v computes vertical edges using the Prewitt filter.

- The function filters.prewitt_h computes horizontal edges using the Prewitt filter.

For example, for vertical edge detection, use prewitt_v and the Python function definition is:

```
from skimage import filters
# The input to filter.prewitt_v has to be a numpy array.
filters.prewitt_v(image)
```

Figure 4.8 is an example of detection of horizontal and vertical edges using the Sobel and Prewitt filters. The vertical Sobel and Prewitt filters have enhanced all the vertical edges, while the corresponding horizontal filters enhanced the horizontal edges and the regular Sobel and Prewitt filters enhanced all edges.

4.3.1.3   Canny Filter

Another popular filter for edge detection is the Canny filter or Canny edge detector [Can86]. This filter uses three parameters to detect edges. The first parameter is the standard deviation, σ, for the Gaussian filter. The second and third parameters are the threshold values, t_1 and t_2. The Canny filter can be best described by the following steps:

1. A Gaussian filter is used on the image for smoothing.

2. An important property of an edge pixel is that it will have a maximum gradient magnitude in the direction of the gradient. So, for each pixel, the magnitude of the gradient given in Equation 4.5 and the corresponding direction, $\theta = \tan^{-1}\left(\dfrac{f_y}{f_x}\right)$, are computed.

3. At the edge points, the first derivative will have either a minimum or a maximum. This implies that the magnitude (absolute value) of the gradient of the image at the edge points is maximum. We will refer to these points as ridge pixels. To identify edge points and suppress others, only ridge tops are retained and other pixels are assigned a value of zero. This process is known as non-maximal suppression.

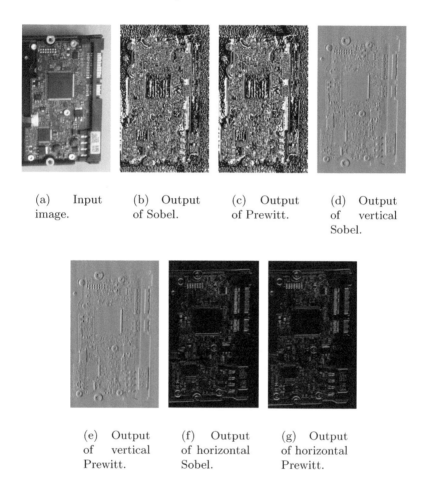

(a) Input image.

(b) Output of Sobel.

(c) Output of Prewitt.

(d) Output of vertical Sobel.

(e) Output of vertical Prewitt.

(f) Output of horizontal Sobel.

(g) Output of horizontal Prewitt.

FIGURE 4.8: Output from vertical, horizontal and regular Sobel and Prewitt filters.

4. Two thresholds, low threshold and high threshold, are then used to threshold the ridges. Ridge pixel values help to classify edge pixels into weak and strong. Ridge pixels with values greater than the high threshold are classified as strong edge pixels, whereas the ridge pixels between low threshold and high threshold are called weak edge pixels.

5. In the last step, the weak edge pixels are 8-connected with strong edge pixels.

The Python function that is used for the Canny filter is:

```
cv2.Canny(image)
Necessary arguments:
input is the input image as an ndarray

Returns: output is an ndarray.
```

The Python code for the Canny filter is given below. The code does not need much explanation.

```
import cv2

# Opening the image.
a = cv2.imread('../Figures/maps1.png')
# Performing Canny edge filter.
b = cv2.Canny(a, 100, 200)
# Saving b.
cv2.imwrite('../Figures/canny_output.png', b)
```

Figure 4.9(a) is a simulated map consisting of names of geographical features of Antarctica. The Canny edge filter is used on this input image to obtain only edges of the letters as shown in Figure 4.9(b). Note that the edge of the characters are clearly marked in the output.

4.3.2 Second Derivative Filters

As the name indicates, in the second derivative filter, the second derivative is computed in order to determine the edges. Since it requires computing the derivative of a derivative image, it is computationally expensive compared to the first derivative filter.

(a) Input image for Canny filter.

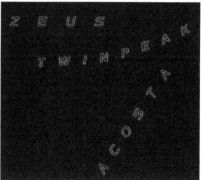

(b) Output of Canny filter.

FIGURE 4.9: Example of Canny filter.

4.3.2.1 Laplacian Filter

One of the most popular second derivative filters is the Laplacian. The Laplacian of a continuous function is given by:

$$\nabla^2 f = \frac{\partial^2 f}{\partial x^2} + \frac{\partial^2 f}{\partial y^2}$$

where $\frac{\partial^2 f}{\partial x^2}$ is the second partial derivative of f in the x direction represents a change of $\frac{\partial f}{\partial x}$ along the horizontal direction and $\frac{\partial^2 f}{\partial y^2}$ is the second partial derivative of f in the y direction represents a change of $\frac{\partial f}{\partial y}$ along the vertical direction. For more details, refer to [Eva10] and [GT01]. The discrete Laplacian used for image processing has several versions. Most widely used Laplacian masks are given in Table 4.9.

TABLE 4.9: Laplacian masks.

0	1	0
−1	4	−1
0	−1	0

−1	−1	−1
−1	8	1
−1	−1	−1

The Python function that is used for the Laplacian along with the arguments is the following:

```
scipy.ndimage.filters.laplace(input, output=None,
  mode='reflect', cval=0.0)

Necessary arguments:
 input is the input image as an ndarray

Optional arguments:
  mode determines the method for handling the array
border by padding. Different options are: constant,
reflect, nearest, mirror, wrap.

  cval is a scalar value specified when the option for
mode is constant. The default value is 0.0.

  origin is a scalar that determines origin of the
filter. The default value 0 corresponds to a filter
whose origin (reference pixel) is at the center. In a
2D case, origin = 0 would mean (0,0).

Returns: output is an ndarray
```

The Python code for the Laplacian filter is given below. The Laplacian is called using the scipy laplace function along with the optional mode for handling array borders.

```
import cv2
import scipy.ndimage

# Opening the image.
```

```
a = cv2.imread('../Figures/imagefor_laplacian.png')
# Performing Laplacian filter.
b = scipy.ndimage.filters.laplace(a,mode='reflect')
cv2.imwrite('../Figures/laplacian_new.png',b)
```

The black-and-white image in Figure 4.10(a) is a segmented CT slice of a human body across the rib cage. The various blobs in the image are the ribs. The Laplacian filter obtained the edges without any artifact.

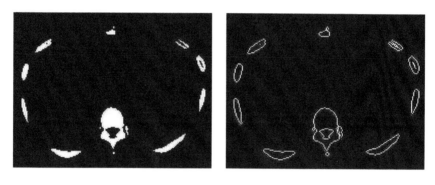

(a) Input image for Laplacian   (b) Output of Laplacian

FIGURE 4.10: Example of the Laplacian filter.

As discussed earlier, a derivative filter adds noise to an image. The effect is magnified when the first derivative image is differentiated again (to obtain a second derivative) as in the case of second derivative filters. Figure 4.11 displays this effect. The image in Figure 4.11(a) is an MRI image from a brain scan. As there are several edges in the input image, the Laplacian filter over-segments the object (creates many edges) as seen in the output, Figure 4.11(b). This results in a noisy image with no discernible edges.

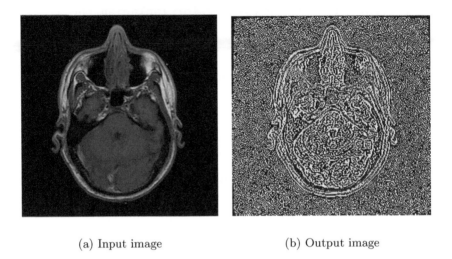

(a) Input image   (b) Output image

FIGURE 4.11: Another example of Laplacian filter.

4.3.2.2   Laplacian of Gaussian Filter

To offset the noise effect from the Laplacian, a smoothing function, Gaussian, is used along with the Laplacian. While the Laplacian calculates the zero-crossing and determines the edges, the Gaussian smooths the noise induced by the second derivative.

The Gaussian function is given by

$$G(r) = -e^{\frac{-r^2}{2\sigma^2}} \tag{4.8}$$

where $r^2 = x^2 + y^2$ and σ is the standard deviation. A convolution of an image with the Gaussian will result in smoothing of the image. The σ determines the magnitude of smoothing. If σ is large then there will be more smoothing, which causes sharp edges to be blurred. Smaller values of σ produce less smoothing.

The Laplacian convolved with Gaussian is known as the Laplacian of Gaussian and is denoted by LoG. Since the Laplacian is the second derivative, the LoG expression can be obtained by finding the second derivative of G with respect to r, which yields

$$\nabla^2 G(r) = -\left(\frac{r^2 - \sigma^2}{\sigma^4}\right) e^{-\frac{r^2}{2\sigma^2}} \qquad (4.9)$$

The LoG mask or filter of size 5-by-5 is given in Table 4.10.

TABLE 4.10: Laplacian of Gaussian mask

0	0	−1	0	0
0	−1	−2	−1	0
−1	−2	16	−2	−1
0	−1	−2	−1	0
0	0	−1	0	0

The following is the Python function for LoG:

```
scipy.ndimage.filters.gaussian_laplace(input,
    sigma, output=None, mode='reflect', cval=0.0)

Necessary arguments:
    input is the input image as an ndarray.

    sigma a floating point value is the standard deviation
    of the Gaussian.

Returns: output is an ndarray
```

The Python code below shows the implementation of the LoG filter. The filter is invoked using the gaussian_laplace function with a sigma of 1.

```
import cv2
import scipy.ndimage

# Opening the image.
a = cv2.imread('../Figures/vhuman_t1.png')
# Performing Laplacian of Gaussian.
```

```
b = scipy.ndimage.filters.gaussian_laplace(a, sigma=1,
    mode='reflect')
cv2.imwrite('../Figures/log_vh1.png', b)
```

Figure 4.12(a) is the input image and Figure 4.12(b) is the output after the application of LoG. The LoG filter was able to determine the edges more accurately compared to the Laplacian alone (Figure 4.11(b)). However, the non-uniform foreground intensity has contributed to formation of blobs (a group of connected pixels).

The major disadvantage of LoG is the computational price as two operations, Gaussian followed by Laplacian, have to be performed. Even though LoG segments the object from the background, it over-segments the edges within the object causing closed loops (also called the spaghetti effect) as shown in the output Figure 4.12(b).

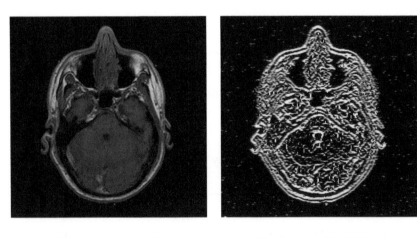

(a) Input image for LoG   (b) Output of LoG filter

FIGURE 4.12: Example of LoG.

4.4   Shape Detecting Filter

4.4.1   Frangi Filter

The Frangi filter [AFFV98] is used to detect vessel-like objects in an image. We will begin the discussion with the fundamental idea of the Frangi filter before discussing the math behind it. Figure 4.4.1 contains two objects. One of the objects is elongated in one direction but not the other, while the second object is almost square. The orthogonal arrows are drawn to be proportional to the length along a given direction. This qualitative geometrical difference can be quantified by finding the eigen values for these two objects. For the elongated object, the eigen value will be larger in the direction of the longer arrow and smaller along the direction of the smaller arrow. On the other hand, for the square object, the eigen value along the direction of the longer arrow is similar to the eigen value along the direction of the smaller arrow. The Frangi filter computes the eigen value on the second derivative (Hessian) image instead of computing the eigen value on the original image.

FIGURE 4.13: Frangi filter illustration.

To reduce noise due to derivatives, the image is smoothed by convolution. Generally, Gaussian smoothing is used. It can be shown that finding the derivative of Gaussian smoothed convolved image is equivalent to finding derivative of a Gaussian convolved with an image. We will determine the second derivative of Gaussian using the formula below where g_σ is a Gaussian.

$$G_\sigma = \begin{bmatrix} \frac{\partial^2 g_\sigma}{\partial x^2} & \frac{\partial^2 g_\sigma}{\partial x \partial y} \\ \frac{\partial^2 g_\sigma}{\partial x \partial y} & \frac{\partial^2 g_\sigma}{\partial y^2} \end{bmatrix} \quad (4.10)$$

We will then determine the local second derivative (Hessian) and its eigen value. For a 2D image, there will be two eigen values (λ_1 and λ_2) for each pixel coordinate. The eigen values are then sorted in increasing order. A pixel is considered to be part of a tubular or vessel-like structure if $\lambda_1 \approx 0$ while $|\lambda_2| > |\lambda_1|$.

For a 3D image, there will be three eigen values (λ_1, λ_2 and λ_3) for each voxel coordinate. The eigen values are then sorted in increasing order. A voxel is considered to be part of a tubular or vessel-like structure if $\lambda_1 \approx 0$ while λ_2 and λ_3 are approximately the same high absolute value and are of the same sign. The bright vessels will have positive values for λ_2 and λ_3 while darker vessels will have negative values for λ_2 and λ_3.

In the code below, we will demonstrate using the Frangi filter programmatically. The image is first opened and converted to grayscale. The image is converted to a numpy array using the np.array function so that it can be fed to the Frangi filter. Finally, a call is made to the Frangi filter located in the skimage.filters module. The output of the frangi function is then saved to a file.

```
import cv2
import numpy as np

import numpy as np
from PIL import Image
from skimage.filters import frangi

img = cv2.imread('../Figures/angiogram1.png')
```

```
img1 = np.asarray(img)
img2 = frangi(img1, black_ridges=True)
img3 = 255*(img2-np.min(img2))/(np.max(img2)-np.min(img2))
cv2.imwrite('../Figures/frangi_output.png', img3)
```

The image in Figure 4.14(a) is the input to the Frangi filter and the image in Figure 4.14(b) is the output of the Frangi filter. The input image is an angiogram that clearly shows multiple vessels enhanced by contrast. The output image contains only pixels that are in the vessel. The contrast of the output image was enhanced for the sake of publication.

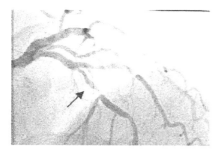
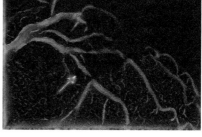

(a) Input image for Frangi filter

(b) Output of Frangi filter (image enhanced for visualization)

FIGURE 4.14: Example of Frangi filter.

4.5 Summary

- The mean filter smooths the image while blurring the edges in the image.

- The median filter is effective in removing salt-and-pepper noise.

- The most widely used first derivative filters are Sobel, Prewitt and Canny.

- Both Laplacian and LoG are popular second derivative filters. The Laplacian is very sensitive to noise. In LoG, the Gaussian smooths the image so that the noise from the Laplacian can be compensated. But LoG suffers from the spaghetti effect.

- The Frangi filter is used for detecting vessel-like structures.

4.6   Exercises

1. Write a Python program to apply a mean filter on an image with salt-and-pepper noise. Describe the output, including the mean filter's ability to remove the noise.

2. Describe how effective the mean filter is in removing salt-and-pepper noise. Based on your understanding of the median filter, can you explain why the mean filter cannot remove salt-and-pepper noise?

3. Can max filter or min filter be used for removing salt-and-pepper noise?

4. Check the scipy documentation available at `http://docs.scipy.org/doc/scipy/reference/ndimage.html`. Identify the Python function that can be used for creating custom filters.

5. Write a Python program to obtain the difference of the Laplacian of Gaussian (LoG). The pseudo code for the program will be as follows:

   (a) Read the image.

(b) Apply the LoG filter assuming a standard deviation of 0.1 and store the image as im1.

(c) Apply the LoG filter assuming a standard deviation of 0.2 and store the image as im2.

(d) Find the difference between the two images and store the resulting image?

6. In this chapter, we have discussed a few spatial filters. Identify two more filters and discuss their properties.

Chapter 5

Image Enhancement

5.1 Introduction

In previous chapters we discussed image filters. The filter enhances the quality of an image so that important details can be visualized and quantified. In this chapter, we discuss a few more image enhancement techniques. These techniques transform the pixel values in the input image to a new value in the output image using a mapping function. We discuss logarithmic transformation, power law transformation, image inverse, histogram equalization, and contrast stretching. For more information on image enhancement refer to [HWJ98],[OR89],[PK81].

5.2 Pixel Transformation

A transformation is a function that maps a set of inputs to a set of outputs so that each input has exactly one output. For example, $T(x) = x^2$ is a transformation that maps inputs to corresponding squares of input. Figure 5.1 illustrates the transformation $T(x) = x^2$ for three inputs.

In the case of images, a transformation takes the pixel intensities of the image as an input and creates a new image where the corresponding pixel intensities are defined by the transformation. Let us consider the transformation, $T(x) = x + 50$. When this transformation is applied

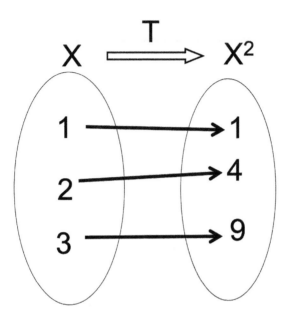

FIGURE 5.1: Illustration of transformation $T(x) = x^2$.

to an image, a value of 50 is added to the intensity of each pixel. The corresponding image is brighter than the input image. Figures 5.2(a) and 5.2(b) are the input and output images of the transformation, $T(x) = x + 50$.

For a grayscale image, the transformation range is given by $[0, L-1]$ where $L = 2^k$ and k is the number of bits in an image. In the case of an 8-bit image, the range is $[0, 2^8 - 1] = [0, 255]$ and for a 16-bit image the range is $[0, 2^{16} - 1] = [0, 65535]$. In this chapter we consider 8-bit grayscale images but the basic principles apply to images of any bit-depth.

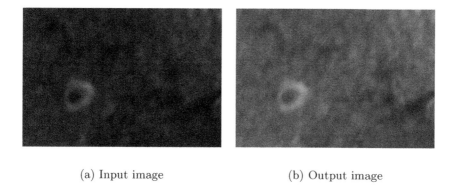

(a) Input image   (b) Output image

FIGURE 5.2: Example of transformation $T(x) = x+50$. Original image reprinted with permission from Mr. Karthik Bharathwaj.

5.3   Image Inverse

Image inverse transformation is a linear transformation. The goal is to transform the dark intensities in the input image to bright intensities in the output image and vice versa. If the range of intensities is $[0, L-1]$ for the input image, then the image inverse transformation at (i, j) is given by the following

$$t(i, j) = L - 1 - I(i, j) \qquad (5.1)$$

where I is the intensity value of the pixel in the input image at (i, j).

For an 8-bit image, the Python code for the image inverse is given below:

```
import cv2



# Opening the image.
im = cv2.imread('../Figures/imageinverse_input.png')
# Performing the inversion operation
```

```
im2 = 255 - im
# Saving the image as imageinverse_output.png in
# Figures folder.
cv2.imwrite('../Figures/imageinverse_output.png', im2)
```

Figure 5.3(a) is a CT image of the region around the heart. Notice that there are several metal objects, bright spots with streaks, emanating in the image. The bright circular object near the bottom edge is a rod placed in the spine, while two arch-shaped metal objects are the valves in the heart. The metal objects are very bright and prevent us from observing other details. The image inverse transformation suppresses the metal objects while enhancing other features of interest such as blood vessels, as shown in Figure 5.3(b).

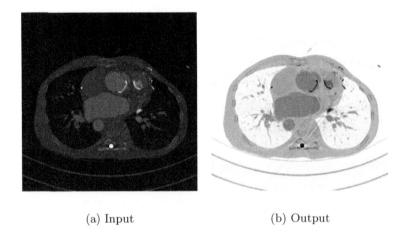

(a) Input   (b) Output

FIGURE 5.3: Example of image inverse transformation. Original image reprinted with permission from Dr. Uma Valeti, Cardiovascular Imaging, University of Minnesota.

5.4 Power Law Transformation

Power law transformation, also known as gamma-correction, is used to enhance the quality of the image. The power transformation at (i, j) is given by

$$t(i, j) = k\, I(i, j)^\gamma \qquad (5.2)$$

where k and γ are positive constants and I is the intensity value of the pixel in the input image at (i, j). In most cases $k = 1$.

If $\gamma = 1$ (Figure 5.4), then the mapping is linear and the output image is the same as the input image. When $\gamma < 1$, a narrow range of dark or low-intensity pixel values in the input image get mapped to a wide range of intensities in the output image, while a wide range of bright or high intensity-pixel values in the input image get mapped to a narrow range of high intensities in the output image. The effect from values of $\gamma > 1$ is opposite that of values $\gamma < 1$. Considering that the intensity range is between $[0, 1]$, Figure 5.4 illustrates the effect of different values of γ for $k = 1$.

The human brain uses gamma-correction to process an image, hence gamma-correction is a built-in feature in devices that display, acquire, or publish images. Computer monitors and television screens have built-in gamma-correction so that the best image contrast is displayed in all the images.

In an 8-bit image, the intensity values range from $[0, 255]$. If the transformation is applied according to Equation 5.2, and for $\gamma > 1$ the output pixel intensities will be out of bounds. To avoid this scenario, in the following Python code the pixel intensities are normalized, $\dfrac{I(i, j)}{max(I)} = I_{norm}$. For $k = 1$, replacing $I(i, j)$ with I_{norm} and then applying the natural log, ln, on both sides of Equation 5.2 will result in

$$\ln(t(i, j)) = \ln(I_{norm})^\gamma = \gamma * \ln(I_{norm}). \qquad (5.3)$$

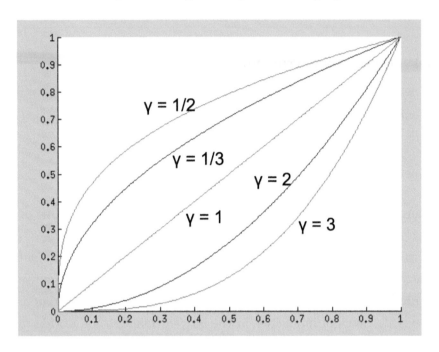

FIGURE 5.4: Graph of power law transformation for different γ.

now, basing both sides by e will give us

$$e^{\ln(t(i,j))} = e^{\gamma * \ln(I_{norm})}. \tag{5.4}$$

Since $e^{\ln(x)} = x$, the left side in the above equation will simplify to

$$t(i,j) = e^{\gamma * \ln(I_{norm})}. \tag{5.5}$$

To have the output in the range $[0, 255]$ we multiply the right side of the above equation by 255 which results in

$$t(i,j) = e^{\gamma * \ln(I_{norm})} * 255. \tag{5.6}$$

This transformation is used in the Python code for power law transformation given below.

```
import cv2
import matplotlib.pyplot as plt
import numpy as np


# Opening the image.
a = cv2.imread('../Figures/angiogram1.png')
# gamma is initialized.
gamma = 0.5
# b is converted to type float.
b1 = a.astype(float)
# Maximum value in b1 is determined.
b3 = np.max(b1)
# b1 is normalized
b2 = b1/b3
# gamma-correction exponent is computed.
b4 = np.log(b2)*gamma
# gamma-correction is performed.
c = np.exp(b4)*255.0
# c is converted to type int.
c1 = c.astype(int)
# Displaying c1
plt.imshow(c1)
```

Figure 5.5(a) is an image of the angiogram of blood vessels. The image is too bright and it is quite difficult to distinguish the blood vessels from background. Figure 5.5(b) is the image after gamma correction with $\gamma = 0.5$; the image is brighter compared to the original image. Figure 5.5(c) is the image after gamma correction with $\gamma = 5$; this image is darker and the blood vessels are visible.

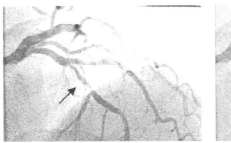
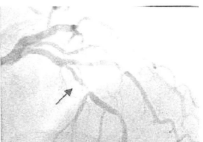

(a) Input image.

(b) Gamma corrected image with $\gamma = 0.5$.

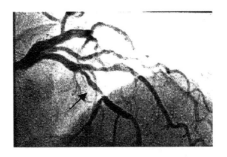

(c) Gamma-corrected image with $\gamma = 5$.

FIGURE 5.5: An example of power law transformation.

5.5 Log Transformation

Log transformation is used to enhance pixel intensities that are otherwise missed due to a wide range of intensity values or lost at the expense of high-intensity values. If the intensities in the image range from $[0, L - 1]$ then the log transformation at (i, j) is given by

$$t(i, j) = k \log(1 + I(i, j)) \qquad (5.7)$$

where $k = \dfrac{L-1}{\log(1+|I_{max}|)}$ and I_{max} is the maximum magnitude value and $I(i,j)$ is the intensity value of the pixel in the input image at (i,j). If both $I(i,j)$ and I_{max} are equal to $L-1$, then $t(i,j) = L-1$. When $I(i,j) = 0$, since $\log(1) = 0$ will give $t(i,j) = 0$. While the end points of the range get mapped to themselves, other input values will be transformed by the above equation. The log can be of any base; however, the common log (log base 10) or natural log (*log* base e) are widely used. The inverse of the above log transformation when the base is e is given by $t^{-1}(x) = e^{\frac{x}{k}} - 1$, which does the opposite of the log transformation.

Similar to the power law transformation with $\gamma < 1$, the log transformation also maps a small range of dark or low-intensity pixel values in the input image to a wide range of intensities in the output image, while a wide range of bright or high-intensity pixel values in the input image get mapped to narrow range of high intensities in the output image. Considering the intensity range is between $[0, 1]$, Figure 5.6 illustrates the log and inverse log transformations.

The Python code for log transformation is given below.

```
import cv2
import numpy, math


# Opening the image.
a = cv2.imread('../Figures/bse.png')
# a is converted to type float.
b1 = a.astype(float)
# Maximum value in b1 is determined.
b2 = numpy.max(b1)
# Performing the log transformation.
c = (255.0*numpy.log(1+b1))/numpy.log(1+b2)
# c is converted to type int.
c1 = c.astype(int)
# Saving c1 as logtransform_output.png.
cv2.imwrite('../Figures/logtransform_output.png', c1)
```

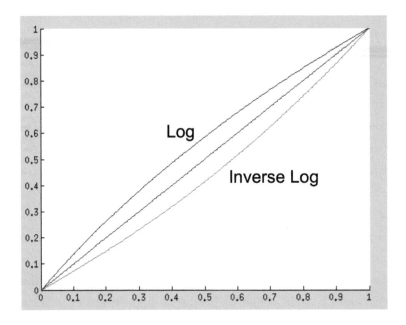

FIGURE 5.6: Graph of log and inverse log transformations.

Figure 5.7(a) is a backscattered electron microscope image. Notice that the image is very dark and the details are not clearly visible. Log transformation is performed to improve the contrast, to obtain the output image shown in Figure 5.7(b).

5.6   Histogram Equalization

The histogram of an image was discussed in Chapter 3, "Image and its Properties." The histogram of an image is a discrete function, its input is the gray-level value and the output is the number of pixels with that gray-level value and can be given as $h(x_n) = y_n$. In a grayscale image, the intensities of the image take values between $[0, L - 1]$. As discussed earlier, low gray-level values in the image (the left side of the

(a) Input (b) Output

FIGURE 5.7: Example of log transformation. Original image reprinted with permission from Mr. Karthik Bharathwaj.

histogram) correspond to dark regions and high gray-level values in the image (the right side of the histogram) correspond to bright regions.

In a low-contrast image, the histogram is narrow, whereas in an image with better contrast, the histogram is spread out. In histogram equalization, the goal is to improve the contrast of an image by rescaling the histogram so that the histogram of the new image is spread out and the pixel intensities range over all possible gray-level values. The rescaling of the histogram will be performed by using a transformation. To ensure that for every gray-level value in the input image there is a corresponding output, a one-to-one transformation is required; that is, every input has a unique output. This means the transformation should be a monotonic function. This will ensure that the transformation is invertible.

Before histogram equalization transformation is defined, the following should be computed:

- The histogram of the input image is normalized so that the range of the normalized histogram is $[0, 1]$.

- Since the image is discrete, the probability of a gray-level value, denoted by $p_x(i)$, is the ratio of the number of pixels with a gray value i to the total number of pixels in the image. This is generally called the probability distribution function (PDF).

- The cumulative distribution function (CDF) is defined as $C(i) = \sum_{j=0}^{i} p_x(j)$, where $0 \le i \le L-1$ and where L is the total number of gray-level values in the image. The $C(i)$ is the sum of all the probabilities of the pixel gray-level values from 0 to i. Note that C is an increasing function.

The histogram equalization transformation can be defined as follows:

$$h(u) = round \left(\frac{C(u) - C_{min}}{1 - C_{min}} * (L - 1) \right) \tag{5.8}$$

where C_{min} is the minimum value in the cumulative distribution. For a grayscale image with range between [0, 255], if $C(u) = C_{min}$ then $h(u) = 0$. If $C(u) = 1$ then $h(u) = 255$. The integer value for the output image is obtained by rounding Equation 5.8.

Let us consider an example to illustrate the probability, CDF, and histogram equalization. Figure 5.8 is an image of size 5 by 5. Let us assume that the gray levels of the image range from [0, 255].

The probabilities, CDF as C for each gray-level value along with the output of histogram equalization transformation, are given in Figure 5.9.

The Python code for histogram equalization is given below. The image is read and a flattened image is calculated. The histogram and the CDF of the flattened image are then computed. The histogram equalization is then performed according to Equation 5.8. The flattened image is then passed through the CDF function and then reshaped to the original image shape.

```
import cv2
import numpy as np
```

32	41	30	41	42
50	35	45	48	34
38	36	40	38	37
41	32	50	37	43
37	38	43	46	45

FIGURE 5.8: An example of a 5-by-5 image.

Gray level value	Probability	CDF as C	h(u)
30	1/25	1/25	0
32	2/25	3/25	22
34	1/25	4/25	32
35	1/25	5/25	43
36	1/25	6/25	53
37	3/25	9/25	85
38	3/25	12/25	117
40	1/25	13/25	128
41	3/25	16/25	160
42	1/25	17/25	170
43	2/25	19/25	191
45	2/25	21/25	212
46	1/25	22/25	223
48	1/25	23/25	234
50	2/25	25/25	255

FIGURE 5.9: Probabilities, CDF, histogram equalization transformation.

```
# Opening the image.
img1 = cv2.imread('../Figures/hequalization_input.png')
# 2D array is converted to a 1D array.
fl = img1.flatten()
# Histogram and the bins of the image are computed.
hist,bins = np.histogram(img1,256,[0,255])
# cumulative distribution function is computed
cdf = hist.cumsum()
# Places where cdf=0 is masked or ignored and
# rest is stored in cdf_m.
cdf_m = np.ma.masked_equal(cdf,0)
# Histogram equalization is performed.
num_cdf_m = (cdf_m - cdf_m.min())*255
den_cdf_m = (cdf_m.max()-cdf_m.min())
cdf_m = num_cdf_m/den_cdf_m
# The masked places in cdf_m are now 0.
cdf = np.ma.filled(cdf_m,0).astype('uint8')
# cdf values are assigned in the flattened array.
im2 = cdf[fl]
# im2 is 1D so we use reshape command to.
#  make it into 2D.
im3 = np.reshape(im2,img1.shape)
# Saving im3 as hequalization_output.png
# in Figures folder
cv2.imwrite('../Figures/hequalization_output.png', im3)
```

An example of histogram equalization is illustrated in Figure 5.10. Figure 5.10(a) is a CT scout image. The histogram and CDF of the input image are given in Figure 5.10(b). The output image after histogram equalization is given in Figure 5.10(c). The histogram and CDF of the output image are given in Figure 5.10(d). Notice that the histogram of the input image is narrow compared to the range [0, 255]. The leads (bright slender wires running from top to bottom of the image) are not clearly visible in the input image. After histogram equalization, the histogram of the output image is spread out over all the values in the range and subsequently the image is brighter and the leads are visible.

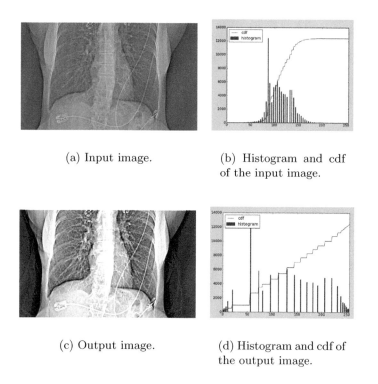

(a) Input image.

(b) Histogram and cdf of the input image.

(c) Output image.

(d) Histogram and cdf of the output image.

FIGURE 5.10: Example of histogram equalization. Original image reprinted with permission from Dr. Uma Valeti, Cardiovascular Imaging, University of Minnesota.

5.7 Contrast Limited Adaptive Histogram Equalization (CLAHE)

In the above histogram equalization method, observe that the output image in 5.10 is too bright. Instead of using the histogram of the whole image, in Contrast Limited Adaptive Histogram Equalization ([Zui94]), the image is divided into small regions and a histogram of each region is computed.

A contrast limit is chosen as a threshold to clip the histogram in each bin, and the pixels above the threshold are not ignored but rather distributed to other bins before histogram equalization is applied.

Let us consider the steps involved:

1. Divide the input image into sub-images of size 8-by-8 (say).

2. Calculate the histogram of each sub-image.

3. Find a PDF as described in Section 5.6.

4. Set a threshold to clip the histograms. Then find the CDF as described in Section 5.6. If the histogram of any bin crosses the clip limit, then the pixels above the clip limit are uniformly distributed to other bins. Since the PDF is clipped, the slope of the CDF will be smaller than the ones in Section 5.6.

5. Apply histogram equalization to each sub-image.

6. Bilinear interpolation is applied to remove artifacts at the boundary of sub-images. We will talk about bilinear interpolation and other interpolation in Chapter 6.

The following is the Python function for the CLAHE filter:

```
from skimage.exposure import equalize_adapthist
```

```
equalize_adapthist(img, clip_limit = 0.02)
```

```
Necessary arguments:
input is the input image as an ndarray.
```

```
Optional arguments:
clip_limit is a floating point number between 0 and 1.
A value close to 1 produces higher contrast.
```

We read the image, 'embryo.png' using cv2. As can be seen in Figure 5.11(a), the contrast of the input image is very poor even though it was enhanced manually for publishing purposes. The image is then passed to the equalize_adapthist function with a clip limit of 0.02. The image is scaled to [0, 255] and saved to a file. The output image is displayed in Figure 5.11(b). As can be seen, the output image contrast is better than the input image. Also more details can be seen in the output image compared to the input image.

```
import cv2
from skimage.exposure import equalize_adapthist

img = cv2.imread('../Figures/embryo.png')
# Applying Clahe.
img2 = equalize_adapthist(img, clip_limit = 0.02)

# Rescaling img2 from 0 to 255.
img3 = img2*255.0
# Saving img3.
cv2.imwrite('../Figures/clahe_output.png', img3)
```

The authors have experienced that CLAHE is particularly useful for image enhancement of MV x-ray images such as seen in radiotherapy.

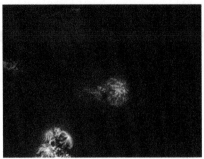

(a) Input (contrast adjusted man-   (b) Output
ually to show details)

FIGURE 5.11: Example of CLAHE.

5.8   Contrast Stretching

Contrast stretching is similar in idea to histogram equalization except that the pixel intensities are rescaled using the pixel values instead of probabilities and CDF. Contrast stretching is used to increase the pixel value range by rescaling the pixel values in the input image. Consider an 8-bit image with a pixel value range of $[a, b]$ where $a > 0$ and $b < 255$. If a is significantly greater than zero and if b is significantly smaller than 255, then the details in the image may not be visible. This problem can be offset by rescaling the pixel value range to $[0, 255]$, a much larger pixel range.

The contrast stretching transformation, $t(i, j)$ is given by the following equation:

$$t(i, j) = 255 * \frac{I(i, j) - a}{b - a} \qquad (5.9)$$

where $I(i, j)$, a, and b are the pixel intensity at (i, j), the minimum pixel value and the maximum pixel value in the input image respectively.

Note that if $a = 0$ and $b = 255$, then there will be no change in pixel intensities between the input and the output images.

The image is read and its minimum and maximum values are computed. The image is converted to float, so that the contrast stretching defined in Equation 5.9 can be performed.

```
import cv2


# Opening the image.
im = cv2.imread('../Figures/hequalization_input.png')
# Finding the maximum and minimum pixel values
b = im.max()
a = im.min()
print(a,b)
# Converting im1 to float.
c = im.astype(float)
# Contrast stretching transformation.
im1 = 255.0*(c-a)/(b-a+0.0000001)
# Saving im2 as contrast_output.png in
# Figures folder
cv2.imwrite('../Figures/contrast_output2.png', im1)
```

In Figure 5.12(a) the minimum pixel value in the image is 7 and the maximum pixel value is 51. After contrast stretching, the output image (Figure 5.12(b)) is brighter and the details are visible.

In Figure 5.13(a), the minimum pixel value in the image is equal to 0 and the maximum pixel value is equal to 255 so the contrast stretching transformation will not have any effect on this image as shown in Figure 5.13(b).

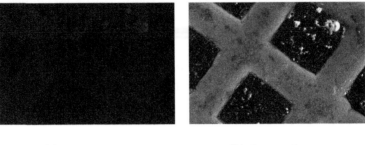

(a) Input image.   (b) Output image.

FIGURE 5.12: An example of contrast stretching where the pixel value range is significantly different from $[0, 255]$.

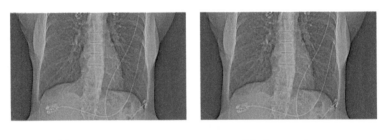

(a) Input image.   (b) Output image.

FIGURE 5.13: An example of contrast stretching where the input pixel value range is same as $[0, 255]$.

5.9 Sigmoid Correction

A sigmoid function is defined as

$$S(x) = \frac{1.0}{1 + e^{-x*gain}} \tag{5.10}$$

The function (Figure 5.10) asymptotically reaches 0 for low negative values or reaches 1 asymptotically for high positive values and is always bound between 0 and 1. In the typical definition of a sigmoid function, the value of gain is 1. However, in the case of sigmoid correction, we will use the gain as a hyper-parameter for fine tuning the image enhancement.

For a gain of 0.5, the slope of the linear region around x value of 0 is smaller than the corresponding slope for a gain of 1. Consequently the saturation of pixel values to either 0 or 1 on either end of the spectrum will happen only for points that are farther away from 0. However, for a gain of 2, the saturation point happens close to x = 0. We can use this property to enhance images.

If we choose a gain of 2, then only pixel values (value along x) close to 0 will retain their pixel values while pixel values farther away from 0 will either be saturated to 0 or 1. Hence only a pixel around 0 will be visible with its gray value range.

Instead, if we choose a gain of 0.5, the pixel values farther away from 0 will retain their gray value range and hence we will visualize a large range of pixel values in the image.

In a scikit image, the sigmoid correction is performed using the formula (Equation 5.11),

$$S(x) = \frac{1.0}{1 + e^{-(cutoff - pixelvalue)*gain}} \tag{5.11}$$

where cutoff is the pixel value around which the sigmoid correction is performed. The pixel values must be normalized to [0, 1] before performing sigmoid correction. The cutoff value is the center value of the pixel around which the gray pixel value range is highlighted in the output image.

The following is the Python function for sigmoid correction.

```
from skimage.exposure import adjust_sigmoid

adjust_sigmoid(img1, gain=15)
```

Necessary arguments:
input is the input image as an ndarray.

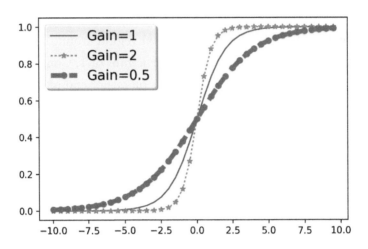

FIGURE 5.14: Effect of gain in a sigmoid function.

```
Optional arguments:
gain is a constant multiplier in exponential's power
of sigmoid function. The default value is 10.
```

In the code below, the image is read and converted to a numpy array. Then sigmoid correction is applied using the "adjust_sigmoid" function with a gain of 15. Since the cutoff is not specified, the default value of 0.5 will be assumed. A gain of 15 will result in a steep slope in the linear region around 0 in Figure 5.14. Thus only the central pixel values will be highlighted and all other pixels farther away from 0 will be set to either 0 or 1.

```
import cv2
from skimage.exposure import adjust_sigmoid

# Reading the image.
img1 = cv2.imread('../Figures/hequalization_input.png')
```

```
# Applying Sigmoid correction.
img2 =  adjust_sigmoid(img1, gain=15)
# Saving img2.
cv2.imwrite('../Figures/sigmoid_output.png', img2)
```

The image in Figure 5.15(a) is sigmoid corrected to produce the output image in Figure 5.15(b). The details of the bones are discernable in the corrected image as opposed to the original image. The choice of the cutoff and gain will determine the quality of the output image.

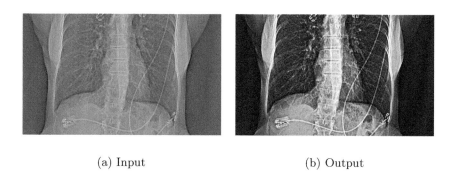

(a) Input   (b) Output

FIGURE 5.15: Example of Sigmoid correction.

5.10   Local Contrast Normalization

Local contrast normalization ([JKRL09]) was developed as part of a computational neural model. The method demonstrates that enhancing the pixel value at a certain location depends only on its neighboring pixels and not the ones farther away from it. The method works by setting the local mean of a pixel to zero and its standard deviation to 1 based on the pixels in the neighborhood.

We begin by creating a difference image (d), computed by finding the difference (Equation 5.12) between the smoothed version of the image and itself. This creates an image whose neighborhood mean is 0. The difference image is then used to compute the standard deviation image (Equation 5.13) after applying a Gaussian smoothing. The final image I_{out} (Equation 5.14) is created by dividing the difference image by the maximum between the local mean of the standard deviation image and the standard deviation image.

$$d = I * \sigma_1 - I \tag{5.12}$$

$$s = \sqrt{d^2 * \sigma_2} \tag{5.13}$$

$$I_{out} = \frac{d}{max(mean_s, s)} \tag{5.14}$$

where I is the original image, σ_1 and σ_2 are the standard deviations for the Gaussian smoothing, * indicates convolution, and $mean_s$ is the mean of the image s.

The convolution operation works on pixels neighboring a given pixel and hence the filter is called a "local contrast normalization."

The image used for this example is a DICOM image and is read using the pydicom module. The image is converted to float and scaled to range [0.0, 1.0].

The "localfilter" function implements the local contrast normalization filter. In the function, the input image is smoothed using a Gaussian. A new image called 'd' is created as a difference between the Gaussian smoothed image and the original image. Since the Gaussian is a weighted mean of the neighborhood pixels, this operation is equivalent to removing the mean from the neighborhood. The mean-corrected image, 'd,' is then squared to obtain the variance and the square root of the variance provides the standard deviation image 's'. A new image max_array is created by finding the maximum between the values in image 's' and the mean value of image 's'. The final image 'y' is created by dividing the image 'd,' which is similar to mean-corrected image and the standard deviation image, 'max_array'.

```python
import pydicom
import numpy as np
import skimage.exposure as imexp
from matplotlib import pyplot as plt
from scipy.ndimage.filters import gaussian_filter
from PIL import Image

def localfilter(im, sigma=(10, 10,)):
    im_gaussian = gaussian_filter(im, sigma=sigma[0])
    d = im_gaussian-im
    s = np.sqrt(gaussian_filter(d*d, sigma=sigma[1]))
    # form an array where all elements have a value of
    mean(s)
    mean_array = np.ones(s.shape)*np.mean(s)
    # find element by element maximum between mean_array
    and s
    max_array = np.maximum(mean_array, s)
    y = d/(max_array+np.spacing(1.0))
    return y


file_name = "../Figures/FluroWithDisplayShutter.dcm"
dfh = pydicom.read_file(file_name, force=True)
im = dfh.pixel_array
# convert to float and scale before applying filter
im = im.astype(np.float)
im1 = im/np.max(im)

sigma = (5, 5,)
im2 = localfilter(im, sigma)
# rescale to 8-bit
im3 = 255*(im2-im2.min())/(im2.max()-im2.min())
```

```
im4 = Image.fromarray(im3).convert("L")
im4.save('../Figures/local_normalization_output.png')
im4.show()
```

The image in Figure 5.16(b) is a local contrast normalized image produced from the input image in Figure 5.16(a). The details of the bones are discernable in the output image as opposed to the original image. In the bright regions outside the anatomy but inside the field of view, the input image is smooth while the corresponding region in the output image is noisy. This is due to the fact that we are forcing regions with low variance (such as smooth regions) and also regions with high variance to have equal variance. The choice of smoothing is a hyper-parameter that needs to be chosen based on the image being processed.

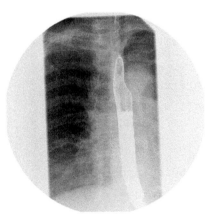 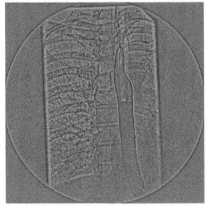

(a) Input (contrast adjusted manually to show details)   (b) Output

FIGURE 5.16: Example of local contrast normalization.

The authors have found that this filter works especially well for highlighting high-contrast objects surrounded by low-contrast structures.

5.11 Summary

- Image inverse transformation is used to invert the pixel intensities in an image. This process is similar to obtaining a negative of a photograph.

- Power law transformation makes the image brighter for $\gamma < 1$ and darker for $\gamma > 1$.

- Log transformation makes the image brighter, while the inverse log makes the image darker.

- Histogram equalization is used to enhance the contrast in an image. In this transformation, a narrow range of intensity values will get mapped to a wide range of intensity values.

- Contrast stretching is used to increase the pixel value range by rescaling the pixel values in the input image.

- Sigmoid correction provides a smooth continuous function for enhancing images around a central cutoff.

- Local contrast normalization enhances the pixel value at a certain location based only on its neighboring pixels and not the ones farther away from it.

5.12 Exercises

1. Explain briefly the need for image enhancement with some examples.

2. Research a few other image enhancement techniques.

3. Consider an image transformation where every pixel value is multiplied by a constant (K). What will be the effect on the image assuming $K < 1$, $K = 1$ and $K > 1$? What will be the impact on the histogram of the output image in relation to the input image?

4. All the transformations discussed in this chapter are scaled from $[0, 1]$. Why?

5. The window or level operation allows us to modify the image, so that all pixel values can be visualized. What is the difference between window or level and image enhancement?

   Clue: One makes a permanent change to the image while the other does not.

6. An image has all pixel values clustered in the lower intensity. The image needs to be enhanced, so that the small range of the low-intensity maps to a larger range. What operation would you use?

7. In sigmoid correction, the choice of the cutoff and gain will determine the quality of the output image. The readers are recommended to try different settings for the hyper-parameter to understand their effect.

8. In local contrast normalization, the choice of the σ_1 and σ_2 affect the outcome. The readers are recommended to try different values to understand their effect.

Chapter 6

Affine Transformation

6.1 Introduction

An affine transformation is a geometric transformation that preserves points, lines and planes. It satisfies the following conditions:

- Collinearity: Points which lie on a line before the transformation continue to lie on the line after the transformation.

- Parallelism: Parallel lines will continue to be parallel after the transformation.

- Convexity: A convex set will continue to be convex after the transformation.

- Ratios of parallel line segments: The ratio of the length of parallel line segments will continue to be the same after transformation.

In this chapter, we will discuss the common affine transformation such as translation, rotation and scaling. We will begin the discussion with the mathematical process to perform affine transformation. We will follow that with specific examples and code for various affine transformations. Finally, we will discuss interpolation which affects the image quality after affine transformation.

6.2   Affine Transformation

The affine transformation is applied as follows:

- Consider every pixel coordinate in the image.

- Calculate the dot product of the pixel coordinate with a transformation matrix. The matrix differs depending on the type of transformation being performed which will be discussed below. The dot product gives the pixel coordinate for the transformed image.

- Determine the pixel value in the transformed image using the pixel coordinate calculated from the previous step. Since the dot product may produce non-integer pixel coordinates, we will apply interpolation (discussed later).

We will discuss the following affine transformation in this chapter. There are other transformations as well but these are most commonly used.

- Translation

- Rotation

- Scaling

6.2.1   Translation

Translation is the process of shifting the image along the various axes (x-, y- and z-axis). For a 2D image, we can perform translation along one or both axes independently. The transformation matrix for translation is defined as:

$$T = \begin{pmatrix} 1 & 0 & 0 \\ 0 & 1 & 0 \\ t_x & t_y & 1 \end{pmatrix} \tag{6.1}$$

If we consider a pixel coordinate (x, y, 1) and perform the dot product with the translation matrix in Equation 6.1, we will obtain the pixel coordinate of the transformed matrix.

$$C_{transformed} = \begin{pmatrix} x & y & 1 \end{pmatrix} \begin{pmatrix} 1 & 0 & 0 \\ 0 & 1 & 0 \\ t_x & t_y & 1 \end{pmatrix} \qquad (6.2)$$

$$C_{transformed} = \begin{pmatrix} x + t_x & y + t_y & 1 \end{pmatrix} \qquad (6.3)$$

Thus every pixel in the transformed image is offset by t_x and t_y along x and y respectively. The value of t_x and t_y may be positive or negative.

The following code implements translation transformation. The image is read and converted in to a numpy array. The transformation matrix is created as an instance of the AffineTransform class. The translation value of (10, 4) is supplied as input to the AffineTransform class. If you need to visualize the value of the transformation matrix similar to one in Equation 6.1, you can print out the content of 'transformation.params'. The transformation is supplied to the warp function which transform the input image img1 to the output image img2.

```
import numpy as np
import scipy.misc, math
from scipy.misc.pilutil import Image
from skimage.transform import AffineTransform, warp

img = Image.open('../Figures/angiogram1.png').convert('L')
img1 = scipy.misc.fromimage(img)

# translate by 10 pixels in x and 4 pixels in y
transformation = AffineTransform(translation=(10, 4))
img2 = warp(img1, transformation)
im4 = scipy.misc.toimage(img2)
```

```
im4.save('../Figures/translate_output.png')
im4.show()
```

The output of the translation is shown below. The image in Figure
6.1(a) is translated to produce the output image in Figure 6.1(b). The
transformed image is translated by 10 pixels to the left and 4 pixels to
the top with reference to the input image.

The missing pixel values on the right and bottom are given a value
of 0 and hence the black pixels on the right and bottom edge. The warp
function's mode parameter can be used to modify this behavior. If the
mode is, "constant" the value of cval parameter to the warp function
will be used instead of a pixel value of 0. If the mode is, "mean",
"median", "maximum", or "minimum" a padding value equal to the
mean, median, maximum or minimum along that vector will be used
respectively. The readers are recommended to read the documentation
for other options. The choice of the padding value affects the quality of
the image and in some cases further computation.

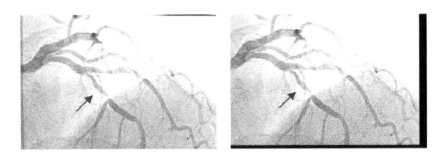

(a) Input image.   (b) Translated image.

FIGURE 6.1: An example of applying translation on an image.

6.2.2 Rotation

Rotation is the process of changing the radial orientation of an image along the various axes with respect to a fixed point. The transformation matrix for a counter-clockwise rotation is defined as:

$$T = \begin{pmatrix} cos(\theta) & sin(\theta) & 0 \\ -sin(\theta) & cos(\theta) & 0 \\ 0 & 0 & 1 \end{pmatrix} \tag{6.4}$$

If we consider a pixel coordinate (x, y, 1) and perform the dot product with the rotation matrix in Equation 6.4, we will obtain the pixel coordinate for the rotated matrix.

$$C_{transformed} = \begin{pmatrix} x & y & 1 \end{pmatrix} \begin{pmatrix} cos(\theta) & sin(\theta) & 0 \\ -sin(\theta) & cos(\theta) & 0 \\ 0 & 0 & 1 \end{pmatrix} \tag{6.5}$$

$$C_{transformed} = \begin{pmatrix} xcos(\theta) - ysin(\theta) & xsin(\theta) + ycos(\theta) & 1 \end{pmatrix} \tag{6.6}$$

The following code implements the rotation transformation. The image is read and converted into a numpy array. The transformation matrix is created as an instance of the AffineTransform class. The rotation value of 0.1 radians is supplied as input to the AffineTransform class. If you need to visualize the value of the transformation matrix similar to the one in Equation 6.4, you can print out the content of 'transformation.params'. The transformation is supplied to the warp function, which transforms the input image img1 to the output image img2 using the transformation.

```
import numpy as np
import scipy.misc, math
from scipy.misc.pilutil import Image
from skimage.transform import AffineTransform, warp
```

```
img = Image.open('../Figures/angiogram1.png').convert('L')
img1 = scipy.misc.fromimage(img)

# rotation angle in radians
transformation = AffineTransform(rotation=0.1)
img2 = warp(img1, transformation)
im4 = scipy.misc.toimage(img2)
im4.save('../Figures/rotate_output.png')
im4.show()
```

The image in Figure 6.2(a) is rotated to produce the output image in Figure 6.2(b). The transformed image is rotated by 0.1 radians with reference to the input image. The missing pixel values on the left and bottom are given a value of 0 that can be modified by supplying appropriate values to the warp function's mode parameter (as discussed in the previous section).

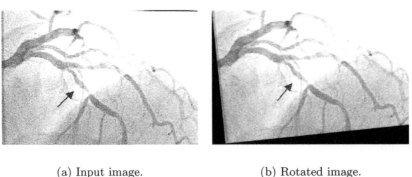

(a) Input image.   (b) Rotated image.

FIGURE 6.2: An example of applying rotation on an image.

6.2.3   Scaling

Scaling is a process of changing the distance (compression or elongation) between points in one or more axes. This change in distance causes the object in the image to appear larger or smaller than the original input. The scaling factor may be different across different axes. The transformation matrix for scaling is defined as:

$$T = \begin{pmatrix} k_x & 0 & 0 \\ 0 & k_y & 0 \\ 0 & 0 & 1 \end{pmatrix} \tag{6.7}$$

If the value of k_x or k_y is less than 1, then the objects in the image will appear smaller ,and missing pixel values will be filled with 0 or based on the value of the warp parameter. If the value of k_x or k_y is greater than 1, then the objects in the image will appear larger. If the value of k_x and k_y are equal, the image is compressed or elongated by the same amount along both axes.

If we consider a pixel coordinate (x, y, 1) and perform the dot product with the scaling matrix in Equation 6.7, we will obtain the pixel coordinate for the scaled matrix.

$$C_{transformed} = \begin{pmatrix} x & y & 1 \end{pmatrix} \begin{pmatrix} k_x & 0 & 0 \\ 0 & k_y & 0 \\ 0 & 0 & 1 \end{pmatrix} \tag{6.8}$$

$$C_{transformed} = \begin{pmatrix} x * k_x & y * k_y & 1 \end{pmatrix} \tag{6.9}$$

The following code implements scaling transformation. The image is read and converted to a numpy array. The transformation matrix is created as an instance of the AffineTransform class. The scaling value of (0.5, 0.5) is supplied as input to the AffineTransform class corresponding to scaling along x and y axes. The transformation is supplied to the warp function, which transforms the input image img1 to the output image img2 using the transformation.

```
import numpy as np
import scipy.misc, math
from scipy.misc.pilutil import Image
```

```
from skimage.transform import AffineTransform, warp

img = Image.open('../Figures/angiogram1.png').convert('L')
img1 = scipy.misc.fromimage(img)

# scale by 1/2 on both x and y.
transformation = AffineTransform(scale=(0.5, 0.5))
img2 = warp(img1, transformation)
im4 = scipy.misc.toimage(img2)
im4.save('../Figures/scale_output.png')
im4.show()
```

The image in Figure 6.3(a) is scaled to produce the output image in Figure 6.3(b). The image is scaled to 0.5 of its original size with reference to the input image along both axes.

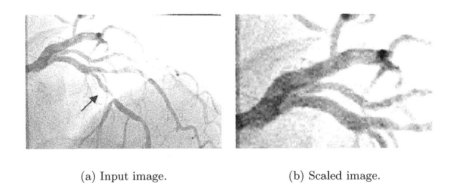

(a) Input image.   (b) Scaled image.

FIGURE 6.3: An example of applying scaling on an image.

6.2.4   Interpolation

To understand the use of interpolation, we will first perform a thought experiment. Consider an image of size 2x2. If this image is scaled to four times its size in all linear dimensions, the new image will be of size 8x8. The original image has only 4 pixel values while the new image needs 64 pixel values. The question is: How can we fill 64 pixels with values given that there are only 4 pixel values? The answer is interpolation.

The various interpolation schemes available are:

1. Nearest-neighbor (order = 0)

2. Bi-linear (order = 1)

3. Bi-quadratic (order = 2)

4. Bi-cubic (order = 3)

5. Bi-quartic (order = 4)

6. Bi-quintic (order = 5)

The order number specified in parentheses is the number used by scikit-image. We will learn about the first 4 interpolation schemes. In all these schemes, the aim is to fill the missing pixel value.

In the nearest-neighbor interpolation,a the missing pixel value is determined based on its immediate neighbors. For a large scaling factor such as 2, we will assign 4 neighbors in the output image to the same pixel value as one of the pixels in the input image, thus making the output image appear pixelated. It is not recommended to use this interpolation even though it is the easiest to implement and also the fastest.

In the bi-linear interpolation, the missing pixel values are determined based on 2x2 pixels around the missing pixels. This results in a smoothed image with fewer artifacts compared to the nearest-neighbor interpolation. Since nearest-neighbor interpolation does not produce a good-quality image compared to other interpolation, it is recommended to use bi-linear at least. In scikit-image, bi-linear is the default interpolation.

In the bi-quadratic interpolation, the missing pixel values are determined based on 3x3 pixels around the missing pixels while in the bi-cubic interpolation, the missing pixel values are determined based on 4x4 pixels around the missing pixels. This results in a smoothed image with fewer artifacts compared to the bi-linear interpolation but at a higher computational cost.

The other two interpolations bi-quartic and bi-quintic, result in smoother interpolation but higher computational cost.

The following code demonstrates the effect of various interpolations. The image is read and converted to a numpy array. The transformation matrix is created as an instance of the AffineTransform class. The scaling value of (0.3, 0.3) is supplied as input to the AffineTransform class corresponding to scaling along x and y axes.

The transformation is then supplied to the warp functio,n which transforms the input image img1 to the output image img2 using the transformation with various interpolation schemes specified using the value for the parameter order. The transformed image for each of the interpolations is then stored.

```python
import numpy as np
import scipy.misc, math
from scipy.misc.pilutil import Image
from skimage.transform import AffineTransform, warp

img = Image.open('../Figures/angiogram1.png').convert('L')
img1 = scipy.misc.fromimage(img)

transformation = AffineTransform(scale=(0.3, 0.3))

# nearest neighbor order = 0
img2 = warp(img1, transformation, order=0)
im4 = scipy.misc.toimage(img2)
im4.save('../Figures/interpolate_nn_output.png')
im4.show()
```

```
# bi-linear order = 1
img2 = warp(img1, transformation, order=1) # default
im4 = scipy.misc.toimage(img2)
im4.save('../Figures/interpolate_bilinear_output.png')
im4.show()

#bi-quadratic order = 2
img2 = warp(img1, transformation, order=2)
im4 = scipy.misc.toimage(img2)
im4.save('../Figures/interpolate_biquadratic_output.png')
im4.show()

#bi-cubic order = 3
img2 = warp(img1, transformation, order=3)
im4 = scipy.misc.toimage(img2)
im4.save('../Figures/interpolate_bicubic_output.png')
im4.show()
```

The image in Figure 6.4(a) is scaled to produce all the other images. The image in Figure 6.4(b) used nearest neighbor interpolation, the image in Figure 6.4(c) used bi-linear interpolation, the image in Figure 6.4(d) used bi-quadratic interpolation and the image in Figure 6.4(e) used bi-cubic interpolation. As can be seen in the image, the nearest-neighbor performed poorly as it exhibits pixelation compared to all other methods. The quality of the image for all other cases are similar but the cost increases significantly for all other methods.

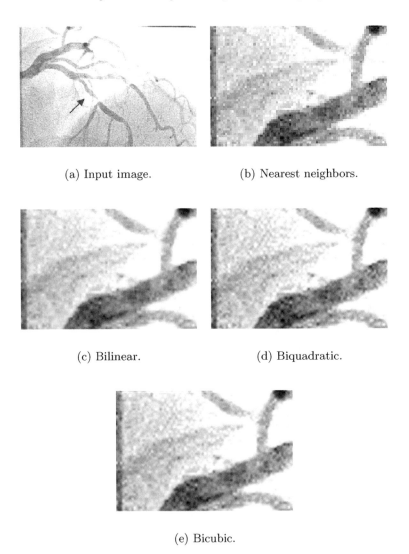

(a) Input image.   (b) Nearest neighbors.

(c) Bilinear.   (d) Biquadratic.

(e) Bicubic.

FIGURE 6.4: An example of applying various interpolation schemes on an image.

6.3   Summary

- Affine transformation is a geometric transformation that preserves points, lines and planes.

- We discussed the commonly used affine transformations such as rotation, translation and scaling.

- We also discussed interpolation, its purpose and the various schemes. It is recommended to not use nearest-neighbor interpolation as it results in pixelation artifact.

6.4 Exercises

1. Consider any of the images used in this chapter. Then, rotate or translate the image by various angles and distance, and for each case, study the histogram. Are the histograms of the input and output image different for different transformations?

2. What happens if you zoom (scale) into the image while keeping the image size the same? Try different zoom levels (2X, 3X, and 4X). Would the histograms of the input image and output image look different?

Chapter 7

Fourier Transform

7.1 Introduction

In Chapter 4, we focused on images in the spatial domain, i.e., the physical world. In this chapter, we will learn about the frequency domain. The process of converting an image from the spatial domain to the frequency domain provides valuable insight into the nature of the image. In some cases, an operation can be performed more efficiently in the frequency domain than in the spatial domain. In this chapter, we introduce the various aspects of the Fourier transform and its properties. We focus exclusively on filtering an image in the frequency domain. Interested readers can refer to [Bra78],[Smi07],[SS03], etc. for more in-depth treatment of Fourier transformation.

The French mathematician Jean Joseph Fourier developed Fourier transforms in an attempt to solve the heat equation. During the process, he recognized that a function can be expressed as infinite sums of sines and cosines of different frequencies, now known as the Fourier series. The Fourier transform is a representation in which any function can be expressed as the integral of sines and cosines multiplied with the weighted function. Also, any function represented in either Fourier series or transform can be reconstructed completely by an inverse process. This is known as inverse Fourier transform.

This result was published in 1822 in the book La Theorie Analitique de la Chaleur. This idea was not welcomed, as at that time mathematicians were interested in and studied regular functions. It took over a century to recognize the importance and power of Fourier series and transforms. Since the development of the fast Fourier transform algorithm, FFT [CT65], the applications of Fourier transforms have affected several fields including remote sensing, signal processing and image processing.

In image processing, Fourier transforms are used for the following:

- Image filtering

- Image compression

- Image enhancement

- Image restoration

- Image analysis

- Image reconstruction

In this chapter we discuss image filtering and enhancement in detail. In Chapter 14, we will discuss use of the Fourier transform in the reconstruction of magnetic resonance images.

7.2   Definition of Fourier Transform

A Fourier transform of a continuous function in one variable $f(x)$ is given by the following equation:

$$F(u) = \int_{-\infty}^{\infty} f(x)e^{-i2\pi ux} dx \qquad (7.1)$$

where $i = \sqrt{-1}$. The function $f(x)$ can be retrieved by finding the inverse Fourier transform of $F(u)$, which is given by the following equation:

$$f(x) = \int_{-\infty}^{\infty} F(u)e^{i2\pi ux}du. \tag{7.2}$$

The Fourier transform of a one-variable discrete function, $f(x)$ for $x = 0, 1, ...L - 1$ is given by the following equation:

$$F(u) = \frac{1}{L} \sum_{x=0}^{L-1} f(x)e^{\frac{-i2\pi ux}{L}} \tag{7.3}$$

for $u = 0, 1, 2, ..., L - 1$. Equation 7.3 is known as the discrete Fourier transform, DFT. Likewise, the inverse discrete Fourier transform, IDFT is given by the following equation:

$$f(x) = \sum_{x=0}^{L-1} F(u)e^{\frac{-i2\pi ux}{L}} \tag{7.4}$$

for $x = 0, 1, 2, ..., L - 1$. Using Euler's formula $e^{i\theta} = \cos\theta + i\,\sin\theta$, the above equation simplifies to

$$F(u) = \frac{1}{L} \sum_{x=0}^{L-1} f(x) \left[\cos\left(\frac{-2ux\pi}{L}\right) - i\sin\left(\frac{-2ux\pi}{L}\right) \right] \tag{7.5}$$

Now, using the fact that the cos is an even function, i.e., $\cos(-\pi) = \cos(\pi)$ and that the sin is an odd function, i.e., $\sin(-\pi) = -\sin(\pi)$, Equation 7.5 can be simplified to:

$$F(u) = \frac{1}{L} \sum_{x=0}^{L-1} f(x) \left[\cos\left(\frac{2ux\pi}{L}\right) + i\sin\left(\frac{2ux\pi}{L}\right) \right] \tag{7.6}$$

$F(u)$ has two parts; the real part constituting the cos is represented as $R(u)$ and the imaginary part constituting the sin is represented as $I(u)$. Each term of F is known as the coefficient of the Fourier transform. Since u plays a key role in determining the frequency of the coefficients of the Fourier transform, u is known as the frequency variable, while x is known as the spatial variable.

Traditionally many experts have compared the Fourier transform to a glass prism. As a glass prism splits or separates the light into various wavelengths or frequencies that form a spectrum, the Fourier transform splits or separates a function into its coefficients, which depend on the frequency. These Fourier coefficients form a Fourier spectrum in the frequency domain.

From Equation 7.6, we know that the Fourier transform is comprised of complex numbers. For computational purposes, it is convenient to represent the Fourier transform in polar form as:

$$F(u) = |F(u)|e^{-i\theta(u)} \qquad (7.7)$$

where $|F(u)| = \sqrt{R^2(u) + I^2(u)}$ is called the magnitude of the Fourier transform and $\theta(u) = \tan^{-1}\left[\frac{I(u)}{R(u)}\right]$ is called the phase angle of the transform. Power, $P(u)$, is defined as the following:

$$P(u) = R^2(u) + I^2(u) = |F(u)|^2. \qquad (7.8)$$

The first value in the discrete Fourier transform is obtained by setting $u = 0$ in Equation 7.3 and then summing the product over all x. Hence, $F(0)$ is nothing but the average of $f(x)$ since $e^0 = 1$. $F(0)$ has a non-zero real part while the imaginary part is zero. Other values of F can be computed in a similar manner.

Let us consider a simple example to illustrate the Fourier transform. Let $f(x)$ be a discrete function with only four values: $f(0) = 2, f(1) = 3, f(2) = 2$ and $f(3) = 1$. Note that the size of f is 4, hence $L = 4$.

$$F(0) = \frac{1}{4}\sum_{x=0}^{3} f(x) = \frac{f(0) + f(1) + f(2) + f(3)}{4} = 2$$

$$F(1) = \frac{1}{4}\sum_{x=0}^{3} f(x)\left[\cos\left(\frac{-2\pi x}{4}\right) - i\sin\left(\frac{-i2\pi x}{4}\right)\right]$$

$$= \frac{1}{4}\left(f(0)\left[\cos\left(\frac{0}{4}\right) + i\sin\left(\frac{0}{4}\right)\right] + f(1)\left[\cos\left(\frac{2\pi}{4}\right) + i\sin\left(\frac{2\pi}{4}\right)\right]\right.$$

$$\left. + f(2)\left[\cos\left(\frac{4\pi}{4}\right) + i\sin\left(\frac{4\pi}{4}\right)\right] + f(3)\left[\cos\left(\frac{6\pi}{4}\right) + i\sin\left(\frac{6\pi}{4}\right)\right]\right)$$

$$= \frac{1}{4}(2(1+0i) + 3(0+1i) + 2(-1+0i) + 1(0-1i))$$

$$= \frac{2i}{4} = \frac{i}{2}$$

Note that $F(1)$ is purely imaginary. For $u = 2$, the value of $F(2) = 0$ and for $u = 3$, the value of $F(3) = \frac{-i}{2}$. The four coefficients of the Fourier transform are $\{2, \frac{i}{2}, 0, \frac{-i}{2}\}$.

7.3   Two-Dimensional Fourier Transform

The Fourier transform for two variables is given by the following equation:

$$F(u, v) = \int_{-\infty}^{\infty}\int_{-\infty}^{\infty} f(x, y)\, e^{-i2\pi(ux+vy)} dx\, dy \qquad (7.9)$$

and the inverse Fourier transform is

$$f(x, y) = \int_{-\infty}^{\infty}\int_{-\infty}^{\infty} F(u, v)e^{i2\pi(ux+vy)} du\, dv. \qquad (7.10)$$

The discrete Fourier transform of a 2D function, $f(x, y)$ with size L and K is given by the following equation:

$$F(u, v) = \frac{1}{LK}\sum_{x=0}^{L-1}\sum_{y=0}^{K-1} f(x, y)e^{-i2\pi\left(\frac{ux}{L} + \frac{vy}{K}\right)} \qquad (7.11)$$

for $u = 1, 2, ..., L - 1$ and $v = 1, 2, ..., K - 1$. Similar to 1D Fourier transform, $f(x, y)$ can be computed from $F(u, v)$ by computing the inverse Fourier transform, given by the following equation:

$$f(x, y) = \sum_{u=0}^{L-1} \sum_{v=0}^{K-1} F(u, v) e^{i2\pi\left(\frac{ux}{L} + \frac{vy}{K}\right)} \tag{7.12}$$

for $x = 1, 2, ..., L - 1$ and $y = 1, 2, ..., K - 1$. As in the case of 1D DFT, u and v are the frequency variables and x and y are the spatial variables. The magnitude of the Fourier transform in 2D is given by the following equation:

$$|F(u, v)| = \sqrt{R^2(u, v) + I^2(u, v)} \tag{7.13}$$

and the phase angle is given by

$$\theta(u, v) = \tan^{-1}\left[\frac{I(u, v)}{R(u, v)}\right] \tag{7.14}$$

and the power is given by

$$P(u, v) = R^2(u, v) + I^2(u, v) = |F(u, v)|^2 \tag{7.15}$$

where $R(u, v)$ and $I(u, v)$ are the real and imaginary parts of the 2D DFT.

The properties of a 2D Fourier transform are:

1. The 2D space with x and y as variables is referred to as the spatial domain and the space with u and v as variables is referred to as the frequency domain.

2. $F(0, 0)$ is the average of all pixel values in the image. It can be obtained by substituting $u = 0$ and $v = 0$ in Equation 7.11. Hence $F(0, 0)$ is the brightest pixel in the Fourier transform image.

3. The two summations are separable. Thus, summation is performed along the x or y-directions first and in the other direction later.

4. The computational complexity of DFT is N^2. Hence a modified method called Fast Fourier Transform (FFT) is used to calculate the Fourier transform. Cooley and Tukey developed the FFT algorithm [CT65]. FFT has a complexity of $NlogN$ and hence the word "Fast" in its name.

7.3.1 Fast Fourier Transform using Python

The following is the Python function for the forward Fast Fourier transform:

```
numpy.fft.fft2(a, s=None, axes=(-2,-1))
```

```
Necessary arguments:
  a is the input image as an ndarray.
```

```
Optional arguments:
s is a tuple of integers that represents the
length of each transformed axis of the output.
The individual elements in s, correspond to
the length of each axis in the input image.
If the length on any axis is less than the
corresponding size in the input image, then
the input image along that axis is cropped. If the
length on any axis is greater than the corresponding
size in the input image, then the input image along
that axis is padded with 0s.
```

```
  axes is an integer used to compute the FFT. If axis
is not specified, the last two axes are used.
```

Returns: output is a complex ndarray.

The Python code for the forward fast Fourier transform is given below.

```
import scipy.fftpack as fftim
from PIL import Image

# Opening the image and converting it to grayscale.
b = Image.open('../Figures/fft1.png').convert('L')
# Performing FFT.
c = abs(fftim.fft2(b))
# Shifting the Fourier frequency image.
d = fftim.fftshift(c)
# Converting the d to floating type and saving it
# as fft1_output.raw in Figures folder.
d.astype('float').tofile('../Figures/fft1_output.raw')
```

In the above code, the image is read and converted to a gray-scale image. The Fast Fourier transform is obtained using the fft2 function and only the absolute value is obtained for visualization. The absolute value image of FFT is then shifted, so that the center of the image is the center of the Fourier spectrum. The center pixel corresponds to a frequency of 0 in both directions. Finally, the shifted image is saved as a raw file for visualization purposes.

The image in Figure 7.1(a) is a slice of Sindbis virus from a transmission electron microscope. The output after performing the FFT is saved as a raw file since the pixel intensities are floating values. ImageJ is used to obtain the logarithm of the raw image and the window level is adjusted to display the corresponding image. Finally, a snapshot of this image is shown in Figure 7.1(b). As discussed previously, the central pixel is the pixel with the highest intensity. This is due to the fact that the average of all pixel values in the original image constitutes the central pixel. The central pixel is $(0, 0)$, the origin. To the left of $(0,0)$ is $-u$ and to the right is $+u$. Similarly, to the top of $(0,0)$ is $+v$ and to the bottom is $-v$. The lower frequency is close to the central pixel and the higher frequency is away from the central pixel.

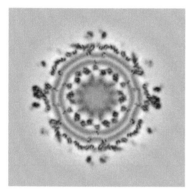
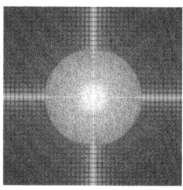

(a) Input for FFT.   (b) Output of FFT.

FIGURE 7.1: An example of 2D Fast Fourier transform. Original image reprinted with permission from Dr. Wei Zhang, University of Minnesota.

The Python function for inverse Fast Fourier transform is given below.

```
numpy.fft.ifft2(a, s=None, axes=(-2,-1))
```

```
Necessary arguments:
  a is a complex ndarray comprising of Fourier
transformed data.
```

```
Optional arguments:


  s is a tuple of integers that represents the length
of each transformed axis of the output. The individual
```

elements in s, correspond to the length of each axis
in the input image. If the length on any axis is less
than the corresponding size in the input image, then
the input image along that axis is cropped. If the
length on any axis is greater than the corresponding
size in the input image, then the input image along
that axis is padded with 0s.

axes is an integer used to compute the FFT. If axis
is not specified, the last two axes are used.

Returns: output is a complex ndarray.

7.4   Convolution

Convolution was briefly discussed in Chapter 4, "Spatial Filters,"
without any mathematical underpinning. In this section, we discuss the
mathematical aspects of convolution.

Convolution is a mathematical operation that expresses the integral
of the overlap between two functions. A simple example is a blurred
image, which is obtained by convolving an un-blurred image with a
blurring function.

There are many cases of blurred images that we see in real life. A
photograph of a car moving at high speed is blurred due to motion.
A photograph of a star obtained from a telescope is blurred by the
particles in the atmosphere. A wide-field microscope image of an object
is blurred by a signal from out-of-plane. Such blurring can be modeled
as a convolution operation and eliminated by the inverse process called
deconvolution.

We begin the discussion with convolution in Fourier space. The convolution operation is expressed mathematically as:

$$[f * g](t) = \int_0^t f(\tau)g(t - \tau)d\tau \qquad (7.16)$$

where f, g are the two functions and the * (asterisk) represents convolution.

The convolution satisfies the following properties:

1. $f * g = g * f$ Commutative Property

2. $f * (g * h) = (f * g) * h$ Associative Property

3. $f * (g + h) = f * g + f * h$ Distributive Property

The convolution operation is simpler in Fourier space than in real space but depending on the size of the image and the functions used, the former can be computationally efficient. In Fourier space, convolution is performed on the whole image at once. However, in spatial domain convolution is performed by sliding the filter window on the image.

7.4.1 Convolution in Fourier Space

Let us assume that the convolution of f and g is the function h.

$$h(t) = [f * g](t). \qquad (7.17)$$

If the Fourier transform of this function is H, then H is defined as

$$H = F.G \qquad (7.18)$$

where F and G are the Fourier transforms of the functions f and g respectively and the . (dot) represents multiplication. Thus, in Fourier space the complex operation of convolution is replaced by a simpler multiplication. The proof of this theorem is beyond the scope of this book. You can find details in most mathematical textbooks on Fourier transform. The formula is applicable irrespective of the number of dimensions of f and g. Hence it can be applied to a 1D signal and also to 3D volume data.

7.5 Filtering in the Frequency Domain

In this section, we discuss applying various filters to an image in the Fourier space. The convolution principle stated in Equation 7.18 will be used for filtering. In lowpass filters, only low frequencies from the Fourier transform are used while high frequencies are blocked. Similarly, in highpass filters, only high frequencies from the Fourier transform are used while the low frequencies are blocked. Lowpass filters are used to smooth the image or reduce noise, whereas highpass filters are used to sharpen edges. In each case, three different filters, namely, ideal, Butterworth and Gaussian, are considered. The three filters differ in the creation of the windows used in filtering.

7.5.1 Ideal Lowpass Filter

The convolution function for the 2D ideal lowpass filter (ILPF) is given by

$$H(u,v) = \begin{cases} 1, & \text{if } d(u,v) \leq d_0 \\ 0, & \text{else} \end{cases} \tag{7.19}$$

where d_0 is a specified quantity and $d(u.v)$ is the Euclidean distance from the point (u,v) to the origin of the Fourier domain. Note that for an image of size M by N, the coordinates of the origin are $\left(\frac{M}{2}, \frac{N}{2}\right)$. So d_0 is the distance of the cutoff frequency from the origin.

For a given image, after the convolution function is defined, the ideal lowpass filter can be performed with element-by-element multiplication of the FFT of the image and the convolution function. Then the inverse FFT is performed on the convolved function to get the output image.

The Python code for the ideal lowpass filter is given below.

```python
import cv2
import numpy, math
import scipy.fftpack as fftim
from PIL import Image


# Opening the image and converting it to grayscale.
b = Image.open('../Figures/fft1.png').convert('L')
# Performing FFT.
c = fftim.fft2(b)
# Shifting the Fourier frequency image.
d = fftim.fftshift(c)


# Intializing variables for convolution function.
M = d.shape[0]
N = d.shape[1]
# H is defined and
# values in H are initialized to 1.
H = numpy.ones((M,N))
center1 = M/2
center2 = N/2
d_0 = 30.0 # cut-off radius


# Defining the convolution function for ILPF.
for i in range(1,M):
    for j in range(1,N):
        r1 = (i-center1)**2+(j-center2)**2
        # Euclidean distance from
        # origin is computed.
        r = math.sqrt(r1)
```

```
            # Using cut-off radius to eliminate
            # high frequency.
            if r > d_0:
                H[i,j] = 0.0
# Converting H to an image.
H =  Image.fromarray(H)
# Performing the convolution.
con = d * H
# Computing the magnitude of the inverse FFT.
e = abs(fftim.ifft2(con))
# Saving e as ilowpass_output.png in
# Figures folder .
cv2.imwrite('../Figures/ilowpass_output.png', e)
```

The image is read and its Fourier transform is determined using the fft2 function. The Fourier spectrum is shifted to the center of the image using the fftshift function. A filter (H) is created by assigning a value of 1 to all pixels within a radius of d_0 and 0 otherwise. Finally, the filter (H) is convolved with the image (d) to obtain the convolved Fourier image (con). This image is inverted using ifft2 to obtain the filtered image in spatial domain. Since high frequencies are blocked, the image 7.2(a) is blurred.

A simple image compression technique can be created using the concept of lowpass filtering. In this technique, all high-frequency data is cleared and only the low-frequency data is stored. This reduces the number of Fourier coefficients stored and consequently needs less storage space on the disk. During the process of displaying the image, an inverse Fourier transform can be obtained to convert the image to the spatial domain. Such an image will suffer from blurring, as high frequency information is not stored. A proper selection of the cut-off radius can reduce the blurring and loss of crucial data in the decompressed image.

7.5.2 Butterworth Lowpass Filter

The convolution function for the Butterworth lowpass filter (BLPF) is given below:

$$H(u, v) = \frac{1}{1 + \left(\frac{d(u,v)}{d_0}\right)^2} \qquad (7.20)$$

where d_0 is the cut-off distance from the origin for the frequency and $d(u, v)$ is the Euclidean distance from the origin. In this filter, unlike the ILPF, the pixel intensity at the cut-off radius does not change rapidly.

The Python code for the Butterworth lowpass filter is given below:

```
import numpy, math
import scipy.fftpack as fftim
from PIL import Image
import cv2


# Opening the image and converting it to grayscale.
b = Image.open('../Figures/fft1.png').convert('L')
 # Performing FFT.
c = fftim.fft2(b)
# Shifting the Fourier frequency image.
d = fftim.fftshift(c)
# Intializing variables for convolution function.
M = d.shape[0]
N = d.shape[1]
# H is defined and
# values in H are initialized to 1.
H = numpy.ones((M,N))
center1 = M/2
center2 = N/2
d_0 = 30.0 # cut-off radius
t1 = 1 # the order of BLPF
t2 = 2*t1
```

```
# Defining the convolution function for BLPF.
for i in range(1,M):
    for j in range(1,N):
        r1 = (i-center1)**2+(j-center2)**2
        # Euclidean distance from
        # origin is computed.
        r = math.sqrt(r1)
        # Using cut-off radius to
        # eliminate high frequency.
        if r > d_0:
            H[i,j] = 1/(1 + (r/d_0)**t1)


# Converting H to an image
H = Image.fromarray(H)
# Performing the convolution.
con = d * H
# Computing the magnitude of the inverse FFT.
e = abs(fftim.ifft2(con))
# Saving e.
cv2.imwrite('../Figures/blowpass_output.png', e)
```

This program is similar to the Python code used for ILPF except for the creation of the filter (H).

7.5.3   Gaussian Lowpass Filter

The convolution function for the Gaussian lowpass filter (GLPF) is given below:

$$H(u, v) = e^{\frac{-d^2(u,v)}{2d_0^2}} \tag{7.21}$$

where d_0 is the cut-off frequency and $d(u, v)$ is the Euclidean distance from origin. The filter creates a much more gradual change in intensity at the cut-off radius compared to the Butterworth lowpass filter.

The Python code for the Gaussian lowpass filter is given below.

```python
import numpy, math
import cv2
import scipy.fftpack as fftim
from PIL import Image


# Opening the image and converting it to grayscale.
b = Image.open('../Figures/fft1.png').convert('L')
# Performing FFT.
c = fftim.fft2(b)
# Shifting the Fourier frequency image.
d = fftim.fftshift(c)
# Intializing variables for convolution function.
M = d.shape[0]
N = d.shape[1]
# H is defined and
# values in H are initialized to 1.
H = numpy.ones((M,N))
center1 = M/2
center2 = N/2
d_0 = 30.0 # cut-off radius
t1 = 2*d_0
# Defining the convolution function for GLPF
for i in range(1,M):
    for j in range(1,N):
        r1 = (i-center1)**2+(j-center2)**2
        # euclidean distance from
        # origin is computed
        r = math.sqrt(r1)
```

```
        # using cut-off radius to
        # eliminate high frequency
        if r > d_0:
            H[i,j] = math.exp(-r**2/t1**2)


# Converting H to an image.
H =  Image.fromarray(H)
# Performing the convolution.
con = d * H
# Computing the magnitude of the inverse FFT.
e = abs(fftim.ifft2(con))
# Saving the image as glowpass_output.png in
# Figures folder .
cv2.imwrite('../Figures/glowpass_output.png', e)
```

Figure 7.1 is the input image to be filtered using ILPF, BLPF and GLPF. The images in Figures 7.2(a), 7.2(b) and 7.2(c) are the outputs of ideal lowpass, Butterworth lowpass, and Gaussian lowpass filters with cut-off radius at 30. Notice how the blurriness varies in the output images. The ILPF is extremely blurred due to the sharp change in the ILPF convolution function at the cut-off radius. There are also severe ringing artifacts, the spaghetti-like structure in the background next to the foreground pixels. In BLPF, the convolution function is continuous which results in less blurring and fewer ringing artifacts compared to ILPF. Since a smoothing operator forms the GLPF convolution function, the output of GLPF is even less blurred when compared to both ILPF and BLPF.

7.5.4   Ideal Highpass Filter

The convolution function for the 2D ideal highpass filter (IHPF) is given by

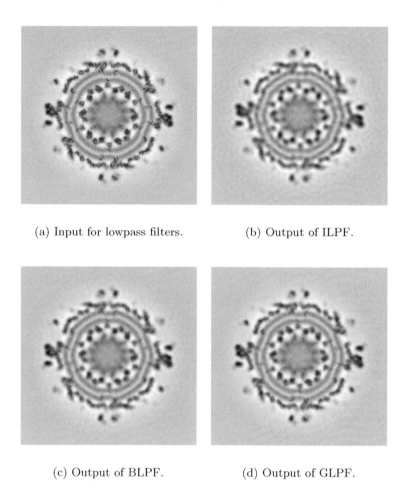

(a) Input for lowpass filters.

(b) Output of ILPF.

(c) Output of BLPF.

(d) Output of GLPF.

FIGURE 7.2: An example of lowpass filters. The input image and all the output images are displayed in the spatial domain.

$$H(u, v) = \begin{cases} 0, & \text{if } d(u, v) \leq d_0 \\ 1, & \text{else} \end{cases} \tag{7.22}$$

where d_0 is the cutoff frequency and $d(u, v)$ is the Euclidean distance from the origin.

The Python code for ideal highpass filter is given below.

```
import cv2
```

```python
import numpy, math
import scipy.fftpack as fftim
from PIL import Image


# Opening the image and converting it to grayscale
a = Image.open('../Figures/endothelium.png').convert('L')
# Performing FFT.
b = fftim.fft2(a)
# shifting the Fourier frequency image
c = fftim.fftshift(b)


# intializing variables for convolution function
M = c.shape[0]
N = c.shape[1]
# H is defined and
# values in H are initialized to 1.
H = numpy.ones((M,N))
center1 = M/2
center2 = N/2
d_0 = 30.0 # cut-off radius


# Defining the convolution function for IHPF.
for i in range(1,M):
    for j in range(1,N):
        r1 = (i-center1)**2+(j-center2)**2
        # Euclidean distance from
# origin is computed.
        r = math.sqrt(r1)
        # Using cut-off radius to
        # eliminate low frequency.
        if 0 < r < d_0:
            H[i,j] = 0.0
# Performing the convolution.
```

```
con = c * H
# Computing the magnitude of the inverse FFT.
d = abs(fftim.ifft2(con))
# Saving the image as ihighpass_output.png in
# Figures folder.
cv2.imwrite('../Figures/ihighpass_output.png', d)
```

In this program, the filter (H) is created by assigning a pixel value of 1 to all pixels above the cut-off radius and 0 otherwise.

7.5.5 Butterworth Highpass Filter

The convolution function for the Butterworth highpass filter (BHPF) is given below:

$$H(u, v) = \frac{1}{1 + \left(\frac{d_0}{d(u,v)}\right)^{2n}} \tag{7.23}$$

where d_0 is the cut-off frequency, $d(u, v)$ is the Euclidean distance from the origin and n is the order of the BHPF.

The Python code for the BHPF is given below.

```
import cv2
import numpy, math
import scipy.misc
import scipy.fftpack as fftim
from PIL import Image


# Opening the image.
a = cv2.imread('../Figures/endothelium.png')
# Converting the image to grayscale.
b = cv2.cvtColor(a, cv2.COLOR_BGR2GRAY)
# Performing FFT.
c = fftim.fft2(b)
# Shifting the Fourier frequency image.
```

```python
d = fftim.fftshift(c)
# Intializing variables for convolution function.
M = d.shape[0]
N = d.shape[1]
# H is defined and
# values in H are initialized to 1.
H = numpy.ones((M,N))
center1 = M/2
center2 = N/2
d_0 = 30.0 # cut-off radius
t1 = 1 # the order of BHPF
t2 = 2*t1


# Defining the convolution function for BHPF.
for i in range(1,M):
    for j in range(1,N):
        r1 = (i-center1)**2+(j-center2)**2
        # Euclidean distance from
        # origin is computed.
        r = math.sqrt(r1)
        # Using cut-off radius to
        # eliminate low frequency.
        if 0 < r < d_0:
            H[i,j] = 1/(1 + (r/d_0)**t2)


# Converting H to an image.
H = Image.fromarray(H)
# performing the convolution
con = d * H
# computing the magnitude of the inverse FFT
e = abs(fftim.ifft2(con))
cv2.imwrite('../Figures/bhighpass_output.png', e)
```

7.5.6 Gaussian Highpass Filter

The convolution function for the Gaussian highpass filter (GHPF) is given below:

$$H(u,v) = 1 - e^{\frac{-d^2(u,v)}{2d_0^2}} \qquad (7.24)$$

where d_0 is the cut-off frequency and $d(u,v)$ the Euclidean distance from the origin.

The Python code for the GHPF is given below.

```python
import cv2
import numpy, math
import scipy.fftpack as fftim
from PIL import Image


# Opening the image and converting it to grayscale.
a = Image.open('../Figures/endothelium.png').convert('L')
# Performing FFT.
b = fftim.fft2(a)
# Shifting the Fourier frequency image.
c = fftim.fftshift(b)


# Intializing variables for convolution function.
M = c.shape[0]
N = c.shape[1]
# H is defined and values in H are initialized to 1.
H = numpy.ones((M,N))
center1 = M/2
center2 = N/2
d_0 = 30.0 # cut-off radius
t1 = 2*d_0
```

```
# Defining the convolution function for GHPF.
for i in range(1,M):
    for j in range(1,N):
        r1 = (i-center1)**2+(j-center2)**2
        # Euclidean distance from
        # origin is computed.
        r = math.sqrt(r1)
        # Using cut-off radius to
        # eliminate low frequency.
        if 0 < r < d_0:
            H[i,j] = 1 - math.exp(-r**2/t1**2)


# Converting H to an image.
H = Image.fromarray(H)
# Performing the convolution.
con = c * H
# Computing the magnitude of the inverse FFT.
e = abs(fftim.ifft2(con))
# Saving the image as ghighpass_output.png in
# Figures folder.
cv2.imwrite('../Figures/ghighpass_output.png', e)
```

The image in Figure 7.3(a) is the endothelium cell. The images in Figures 7.3(b), 7.3(c) and 7.3(d) are the outputs of the IHPF, BHPF and GHPF with cut-off radius at 30. Highpass filters are used to determine edges. Notice how the edges are formed in each case.

7.5.7   Bandpass Filter

A bandpass filter, as the name indicates, allows frequency from a band or range of values. All the frequencies from outside the band are set to zero. Similar to the lowpass and highpass filters, bandpass

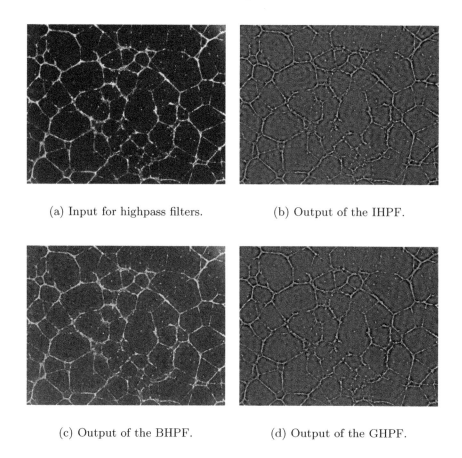

(a) Input for highpass filters.　　　(b) Output of the IHPF.

(c) Output of the BHPF.　　　(d) Output of the GHPF.

FIGURE 7.3: An example of highpass filters. The input image and all the output images are displayed in the spatial domain.

filters can be Ideal, Butterworth or Gaussian. Let us consider the ideal bandpass filter, IBPF.

The Python code for the IBPF is given below.

```
import scipy.misc
import numpy, math
import scipy.fftpack as fftim
from PIL import Image
import cv2
```

```python
# Opening the image and converting it to grayscale.
b = Image.open('../Figures/fft1.png').convert('L')
# Performing FFT.
c = fftim.fft2(b)
# Shifting the Fourier frequency image .
d = fftim.fftshift(c)
# Intializing variables for convolution function.
M = d.shape[0]
N = d.shape[1]
# H is defined and
# values in H are initialized to 1.
H = numpy.zeros((M,N))
center1 = M/2
center2 = N/2
d_0 = 30.0 # minimum cut-off radius
d_1 = 50.0 # maximum cut-off radius


# Defining the convolution function for bandpass
for i in range(1,M):
    for j in range(1,N):
        r1 = (i-center1)**2+(j-center2)**2
        # Euclidean distance from
        # origin is computed.
        r = math.sqrt(r1)
        # Using min and max cut-off to create
# the band or annulus.
        if r > d_0 and r < d_1:
            H[i,j] = 1.0


# Converting H to an image.
H = Image.fromarray(H)
# Performing the convolution.
con = d * H
```

```
# Computing the magnitude of the inverse FFT.
e = abs(fftim.ifft2(con))
# Saving the image as ibandpass_output.png.
cv2.imwrite('../Figures/ibandpass_output.png', e)
```

The difference between this program compared to highpass or low-pass filters is in creation of the filter. In the bandpass filter, the minimum cut-off radius is set to 30 and the maximum cut-off radius is set to 50. Only intensities between 30 and 50 are passed and everything else is set to zero. Figure 7.4(a) is the input image and Figure 7.4(b) is the output image for the IBPF. Notice that the edges in the output image of the IBPF is sharp compared to the input. Similar filters can be created for Butterworth and Gaussian filters using the formula discussed earlier.

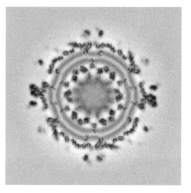

(a) Input of the IBPF.   (b) Output of the IBPF.

FIGURE 7.4: An example of IBPF. The input and the output are displayed in the spatial domain.

7.6 Summary

- Lowpass filters are used for noise reduction or smoothing. Highpass filters are used for edge enhancement or sharpening.

- In ideal lowpass and highpass filters, Butterworth and Gaussian were considered.

- A bandpass filter has minimum cut-off and maximum cut-off radii.

- Convolution can be viewed as the process of combining two images. Convolution is multiplication in the Fourier domain. The inverse process is called deconvolution.

- The Fourier transform can be used for image filtering, compression, enhancement, restoration and analysis.

7.7 Exercises

1. The Fourier transform is one method for converting any function as a sum of basis functions. Perform research and find at least two other such methods. Write a report on their use in image processing.

   Hint: Wavelet, z-transform

2. An example for determining the Fourier coefficient was shown earlier. However, the discussion was limited to 4 coefficients. Determine the 5th coefficient assuming $f(4) = 2$.

3. The central pixel in the Fourier image is brighter compared to other pixels. Why?

4. The image in Figure 7.2(b) has a fuzzy structure next to the object. What is this called? What causes the artifact? Why are there fewer artifacts in BLPF and GLPF output images?

5. Consider an image of size 10,000-by-10,000 pixels that needs to be convolved with a filter of size 100-by-100. Comment about the most efficient method for convolving. Would it be convolution in the spatial domain or Fourier?

Chapter 8

Segmentation

8.1 Introduction

Segmentation is the process of separating an image into multiple logical regions. The regions can be defined as pixels sharing similar characteristics such as intensity, texture, etc. There are many methods of segmentation. They can be classified as follows:

- Histogram-based segmentation

- Region-based segmentation

- Edge segmentation

- Differential equation-based methods

- Contour methods

- Graph partitioning methods

- Model based segmentation

- Clustering methods, etc.

In this chapter, we discuss histogram and region-based and contour segmentation methods. Edge-based segmentation was discussed in Chapter 4, "Spatial Filters." The other methods are beyond the scope of this book. Interested readers can refer to [GWE09],[Rus11] and [SHB+99] for more details.

8.2 Histogram-Based Segmentation

In the histogram-based method (Figure 8.1), a threshold is determined by using the histogram of the image. Each pixel in the image is compared with the threshold value. If the pixel intensity is less than the threshold value, then the corresponding pixel in the segmented image is assigned a value of zero. If the pixel intensity is greater than the threshold value, then the corresponding pixel in the segmented image is assigned a value of 1. Thus,

if $pv \geq threshold$ **then**
  $segpv = 1$
else
  $segpv = 0$
end if

where pv is the pixel value in the image, $segpv$ is the pixel value in the segmented image.

The various histogram-based methods differ in their techniques of determining the threshold. We will discuss Otsu's method and the Renyi entropy method. In images with a non-uniform background, a global threshold value from the histogram-based method might not be optimal. In such cases, local adaptive thresholding (discussed later) may be used.

8.2.1 Otsu's Method

Otsu's method [Ots79] works best if the histogram of the image is bi-modal, but can be applied to other histograms as well. A bi-modal histogram is a type of histogram (similar to Figure 8.1) containing two distinct peaks separated by a valley. One peak is the background and the other is the foreground. Otsu's algorithm searches for a threshold

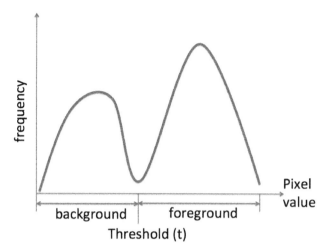

FIGURE 8.1: Threshold divides the pixels into foreground and background.

value that maximizes the variance between the two groups foreground and background, so that the threshold value can better segment the foreground from the background.

Let L be the number of intensities in the image. For an 8-bit image, $L = 2^8 = 256$. For a threshold value, t, the probabilities, p_i, of each intensity are calculated. Then the probability of the background pixels is given by $P_b(t) = \sum_{i=0}^{t} p_i$ and the probability of foreground pixels is given by $P_f(t) = \sum_{i=t+1}^{L-1} p_i$. Let $m_b = \sum_{i=0}^{t} ip_i$, $m_f = \sum_{i=t+1}^{L-1} ip_i$ and $m = \sum_{i=0}^{L-1} ip_i$ represent the average intensities of the background, the foreground, and the whole image, respectively. Let v_b, v_f and v be the variance of the background, foreground, and the whole image, respectively. Then the variance within the groups is given by Equation 8.1 and the variance in between the groups is given by Equation 8.2.

$$v_{within} = P_b(t)v_b + P_f(t)v_f \tag{8.1}$$

$$v_{inbetween} = v - v_{within} = P_b P_f (m_b - m_f)^2. \tag{8.2}$$

For different threshold values, this process of finding variance within the groups and variance between the groups is repeated. The threshold value that maximizes the variance between the groups or minimizes the variance within the group is considered Otsu's threshold. All pixel values with intensities less than the threshold value are assigned a value of zero and all pixel values with intensities greater than the threshold value are assigned a value of one.

In the case of a color image, since there are three channels, Red, Green, and Blue, a different threshold value for each channel is calculated.

The following is the Python function for Otsu's method:

```
skimage.filter.threshold_otsu(image, nbins=256)

Description of function arguments:

necessary argument:
image = input image in gray-scale

optional argument:
nbins = number of bins that should be considered
to calculate the histogram.
```

The Python code for Otsu's method is given below.

```
import cv2
import numpy
from PIL import Image
from skimage.filters.thresholding import threshold_otsu
```

```
# Opening the image and converting it to grayscale
a = Image.open('../Figures/sem3.png').convert('L')
a = numpy.asarray(a)
thresh = threshold_otsu(a)
# Pixels with intensity greater than the
# "threshold" are kept.
b = 255*(a > thresh)
# Saving the image.
cv2.imwrite('../Figures/otsu_output.png', b)
```

In Figure 8.2(a) is a scattered electron image of an atomic element in two different phases. We segment the image using Otsu's method. The output is given in Figure 8.2(b).

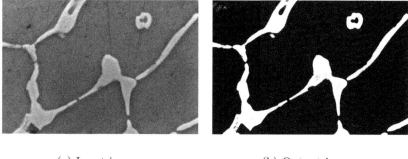

(a) Input image.   (b) Output image.

FIGURE 8.2: An example of Otsu's method. Original image reprinted with permission from Karthik Bharathwaj.

Otsu's method uses the histogram to determine the threshold and hence is very much dependent on the image pixel values. Figure 8.3(a) is an image of a spin-wheel. Otsu's method is used to segment this image, and the segmented output image is shown in Figure 8.3(b). Due to the shadow on the wheel in the input image, Otsu's method did not segment the spin-wheel accurately.

(a) Input image for Otsu's method. 　　　(b) Output of Otsu's method.

FIGURE 8.3: Another example of Otsu's method.

8.2.2 Renyi Entropy

Renyi entropy-based segmentation is very useful when the object of interest is small compared to the whole image i.e., the threshold is at the right tail of the histogram. For example, in the CT image of an abdomen shown in Figure 8.4(b), the tissue and background occupy more area in comparison to the bone. In the histogram, the background and tissue pixels have low pixel intensity while the bone region has high intensity.

In information theory and image processing, entropy quantifies the uncertainty or randomness of a variable. This concept was first introduced by Claude E. Shannon in his 1948 paper "A Mathematical Theory of Communication" [Sha48]. This paper launched Shannon as the father of information theory. In information theory and also in image processing, entropy is measured in bits where each pixel value is considered as an independent random variable.

Shannon entropy is given by

$$H_1(x) = -\sum_{i=1}^{n} p(x_i) \log_a(p(x_i)) \tag{8.3}$$

where x_i is the random variable with $i = 1, 2, ..., n$ and $p(x_i)$ is the probability of the random variable x_i and the base a can be 2, e, or 10.

Alfred Renyi, a Hungarian mathematician, introduced Renyi entropy in his paper [Ren61] in 1961. Renyi entropy is a generalization of Shannon entropy and many other entropies and is given by the following equation:

$$H_\alpha(x) = \frac{1}{1-\alpha} \log_a \left(\sum_{i=1}^{n} (p(x_i))^\alpha \right) \tag{8.4}$$

where x_i is the random variable with $i = 1, 2, ..., n$ and $p(x_i)$ is the probability of the random variable x_i and the base a can be 2, e or 10. Renyi entropy equals Shannon entropy for $\alpha \to 1$.

The histogram of the image is used as an independent random variable to determine the threshold. The histogram is normalized by dividing each frequency with the total number of pixels in the image. This will ensure that the sum of the frequencies after normalization is one. This is the probability distribution function (pdf) of the histogram. The Renyi entropy can then be calculated for this pdf.

The Renyi entropy is calculated for all pixels below and above the threshold. These will be referred to as background entropy and foreground entropy respectively. This process is repeated for all the pixel values in the pdf. The total entropy is calculated as the sum of background entropy and foreground entropy for each pixel value in the pdf. The graph of the total entropy has one absolute maximum. The threshold value corresponding to that absolute maximum is the threshold (t) for segmentation.

The following is the Python code for Renyi entropy for an 8-bit (grayscale) image. The program execution begins with opening the CT image. The image is then processed by the function renyi_seg_fn. The function obtains the histogram of the image and calculates the pdf by dividing each histogram value by the total number of pixels. Two arrays, h1 and h2, are created to store the background and foreground Renyi entropy. For various thresholds, the background and foreground Renyi entropy are calculated using Equation 8.4. The total entropy is the sum of the background and foreground Renyi entropies. The threshold value for which the entropy is maximum is the Renyi entropy threshold.

```python
import cv2
from PIL import Image
import numpy as np
import skimage.exposure as imexp
import matplotlib.pyplot as plt


# Defining function
def renyi_seg_fn(im, alpha):
    hist, _ = imexp.histogram(im)
    # Convert all values to float
    hist_float = np.array([float(i) for i in hist])
    # compute the pdf
    pdf = hist_float/np.sum(hist_float)
    # compute the cdf
    cumsum_pdf = np.cumsum(pdf)
    s, e = im.min(), im.max()
    scalar = 1.0/(1.0-alpha)
    # A very small value to prevent error due to log(0).
    eps = np.spacing(1)

    rr = e-s
    # The inner parentheses is needed because
    # the parameters are tuple.
    h1 = np.zeros((rr, 1))
    h2 = np.zeros((rr, 1))
    # The following loop computes h1 and h2
    # values used to compute the entropy.
    for ii in range(1, rr):
        iidash = ii+s
        temp0 = pdf[0:iidash]/(cumsum_pdf[iidash])
        temp1 = np.power(temp0, alpha)
        h1[ii] = np.log(np.sum(temp1)+eps)
        temp0 = pdf[iidash+1:e]/(1.0-cumsum_pdf[iidash])
```

```
    temp2 = np.power(temp0, alpha)
    h2[ii] = np.log(np.sum(temp2)+eps)


T = h1+h2
# Entropy value is calculated
T = T*scalar
T = T.reshape((rr, 1))[:-2]
# location where the maximum entropy
# occurs is the threshold for the renyi entropy
thresh = T.argmax(axis=0)
return thresh
```

```
# Main program
# Opening the image and converting it to grayscale.
a = Image.open('../Figures/CT.png').convert('L')
a = np.array(a)
# Computing the threshold by calling the function.
thresh = renyi_seg_fn(a, 3)
print('The renyi threshold is: ', thresh[0])
b = 255*(a > thresh)
# Saving the image as renyi_output.png
cv2.imwrite('../Figures/renyi_output.png', b)
```

Figure 8.4(a) is a CT image of the abdomen. The histogram of this image is given in Figure 8.4(b). Notice that the bone region (higher pixel intensity) is on the right side of the histogram and are fewer in number compared to the whole image. Renyi entropy is performed on this image to segment the bone region alone. The segmented output image is given in Figure 8.4(c).

For more details on thresholding, refer to [Par91], [SSW88] and [SPK98].

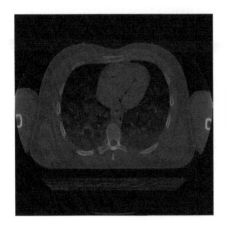

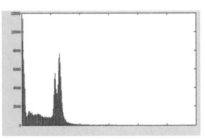

(a) Input image.   (b) Histogram of the input.

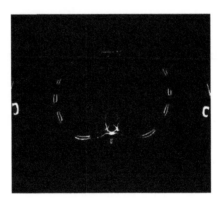

(c) Output image.

FIGURE 8.4: An example of Renyi entropy.

8.2.3   Adaptive Thresholding

As we have seen in Section 8.2.1, Otsu's method, a global threshold, might not provide accurate segmentation. Adaptive thresholding helps solve this problem. In adaptive thresholding, the image is first divided into many sub-images. The threshold value for each sub-image is computed and is used to segment the sub-image. The threshold value for the sub-image can be computed using the mean or median or Gaussian methods. In the case of the mean method, the mean of the sub-image is used as a threshold, while for the median method, the median of the sub-image is used as a threshold. A custom formula can also be used to compute the threshold, for example, we can use an average of maximum and minimum pixel values in the sub-image. By appropriate programming, any of the histogram-based segmentation methods can be converted into an adaptive thresholding method.

The following is the Python function for adaptive thresholding:

```
cv2.AdaptiveThreshold(image, dst, maxValue,
    adaptiveMethod, thresholdType, blockSize, C)

Necessary arguments:
image is a gray-scale image of type numpy array.


dst is the thresholded image as an ndarray.


maxValue is the maximum pixel value in the image.


C is a constant value that should be subtracted
from every pixel value (see below).


blockSize is an odd integer that specifies the size
of the adaptive thresholding window.


adaptiveMethod can be mean or Gaussian. For the
mean method, the threshold is calculated as the mean
of the pixel value within the blockSize minus the
parameter C. For the Gaussian method, the threshold
is calculated as the weighted sum of the region within
the blockSize minus the parameter C.


thresholdType can be either THRESH_BINARY or
```

THRESH_BINARY_INV.   In the former, if a given
pixel value is greater than the threshold, then that pixel
in the output image will be set to a maximum value and
other pixels will be set to zero.
In the latter, if a given pixel value is smaller than the
threshold, then that pixel in the output image will be set
to a maximum value other pixels will be set to zero.

Returns: output is a thresholded image as an ndarray.

The Python code for adaptive thresholding is given below.

```
import cv2
import numpy
from PIL import Image
from skimage.filters import threshold_local


# Opening the image and converting it to grayscale.
a = Image.open('../Figures/adaptive_example1.png').
        convert('L')
a = numpy.asarray(a)
# Performing adaptive thresholding.
b = cv2.adaptiveThreshold(a,a.max(), cv2.ADAPTIVE_THRESH_
        MEAN_C, cv2.THRESH_BINARY,21,10)
# Saving the image as adaptive_output.png
# in the folder Figures.
cv2.imwrite('../Figures/adaptive_output.png', b)
```

In the above code, adaptive thresholding is performed using blocks
of size 40-by-40. The parameter C is set to 10. The method used is
mean thresholding. The image in Figure 8.5(a) is the input image. The
light region is non-uniform and it varies from dark on the left edge to
bright on the right edge. Otsu's method uses a single threshold for the

entire image and hence does not segment the image properly (Figure 8.5(b)). The text in the left section of the image is obscured by the dark region. The adaptive thresholding method (Figure 8.5(c)) uses a local threshold and segments the image accurately.

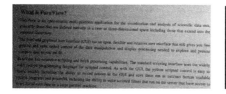 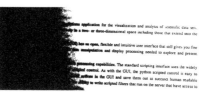

        (a) Input image.   (b) Output using Otsu's method.

(c) Output using adaptive thresholding.

FIGURE 8.5: An example of thresholding with adaptive vs. Otsu thresholding.

8.3 Region-Based Segmentation

A region is a group or collection of pixels that have similar properties sharing the same characteristics. The characteristics can be pixel intensities, texture, or some other physical feature.

Previously, we have used a threshold obtained from a histogram to segment the image. In this section we demonstrate techniques that are based on the region of interest. In Figure 8.3, the objects are labeled as R_1, R_2, R_3, R_4 and the background as R_5.

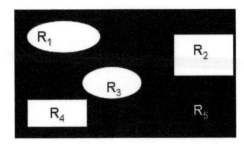

FIGURE 8.6: An example of an image for region-based segmentation.

The different regions constitute the image, $\bigcup_{i=1}^{5} R_i = I$ where I represents the whole image. No two regions overlap, $R_i \cap R_j = \emptyset$ for $i \neq j$. Every region is connected, with I representing the image and R_i representing the regions for $i = 1$ to n. We can now formulate basic rules that govern the region-based segmentation.

1. All the regions combined should equal the image, $\bigcup_{i=1}^{n} R_i = I$.

2. Each region, R_i is connected for $i = 1$ to n.

3. No two regions overlap, $R_i \cap R_j = \emptyset$.

To segment the regions, we need some a priori information. This a priori information is the seed pixels, pixels that are part of the foreground. The seed pixels grow by considering the pixels in their neighborhood that have similar properties. This process connects all the pixels in a region with similar properties. The region-growing process will terminate when there are no more pixels to add that share the same characteristics of the region.

It might not always be possible to have a priori knowledge of the seed pixels. In such cases, a list of characteristics of different regions should be considered. Then pixels that satisfy the characteristics of a particular region will be grouped together. The most popular region-based segmentation method is watershed segmentation.

8.3.1 Watershed Segmentation

To perform watershed segmentation, let's consider a grayscale image as an example. The grayscale values of the image represent the peaks and valleys of the topographic terrain of the image. The lowest valley in an object is the absolute minimum. The highest grayscale value corresponds to the highest point in the terrain. The watershed segmentation can be explained as follows: all the points in a region where, if a drop of water was place,d will settle to the absolute minimum are known as the catchment basin of that minimum or watershed. If water is supplied at a uniform rate from the absolute minimum in an object, as water fills up the object, at some point water will overflow into other objects. Dams are constructed to stop water from overflowing into other objects/regions. These dams are the watershed segmentation lines. The watershed segmentation lines are edges that separate one object from another.

Now let us look at how the dams are constructed. For simplicity, let us assume that there are two regions. Let R_1 and R_2 be two regions and let C_1 and C_2 be the corresponding catchment basins. Now for each time step, the regions that constitute the catchment basins are increased. This can be achieved by dilating the regions with a structuring element of size 3-by-3 (say). If C_1 and C_2 become one connected region in the time step n, then at the time step $n - 1$ the regions C_1 and C_2 were disconnected. The dams or the watershed lines can be obtained by taking the difference of images at time steps n and $n - 1$.

In 1992, F. Meyer proposed an algorithm to segment color images, [Mey92], [Mey94]. Internally, cv2.watershed uses Meyer's flooding algorithm to perform watershed segmentation. Meyer's algorithm is outlined below:

1. The original input image and the marker image are given as inputs.

2. For each region in the marker image, its neighboring pixels are placed in a ranked list according to their gray levels.

3. The pixel with the highest rank (highest gray level) is compared with the labeled region. If the pixels in the labeled region have the same gray level as the given pixel, then the pixel is included in the labeled region. Then a new ranked list with the neighbors is formed. This step contributes to the growth of the labeled region.

4. The above step is repeated until there are no elements in the list.

Prior to performing watershed segmentation, the image has to be preprocessed to obtain a marker image. Since the water is supplied from catchment basins, these basin points are guaranteed foreground pixels. The guaranteed foreground pixel image is known as the marker image.

The preprocessing operations that should be performed before watershed segmentation are as follows:

1. Foreground pixels are segmented from the background pixels.

2. Erosion is performed to obtain foreground pixels only. Erosion is a morphological operation in which the background pixels grow and foreground pixels shrink. Erosion is explained in detail in Chapter 9, "Morphological Operations" in Sections 9.4 and 9.7.

3. Distance transform creates an image where every pixel contains the value of the distance between itself and the nearest background pixel. Thresholding is done to obtain the pixels that are farthest away from the background pixels and are guaranteed to be foreground pixels.

4. All the connected pixels in a region are given a value in the process known as labeling. The labeled image is used as a marker image. Further explanation of labeling can be found in Chapter 10, "Image Measurements" in Section 10.2.

These operations along with the watershed are used in the cv2.watershed code provided below.

All the cv2 functions that are used for preprocessing such as erode, threshold, distance transform, and watershed are explained below. More detailed documentation can be found at [Ope20a]. This will be followed by the Python program using cv2 module.

The cv2 function for erosion is as follows:

```
cv2.erode(input, element, iterations, anchor,
                borderType, borderValue)


Necessary arguments:


input is the input image.


iterations is an integer value corresponding to
the number of times erosion is performed.


Optional arguments:


element is the structuring element. The default
value is None. If element is specified, then anchor
is the center of the element. The default value
is (-1,-1).


borderType is similar to mode argument in convolve
function. If borderType is constant then borderValue
should be
specified.


Returns: An eroded image.
```

The cv2 function for thresholding is given below:

```
cv2.threshold(input, thresh, maxval, type)
```

Necessary arguments:

input is an input array. It can be either 8 bit
or 32 bit.

thresh is the threshold value.

Optional arguments:

maxval should be assigned and will be used when the
threshold type is THRESH_BINARY or THRESH_BINARY_INV.
This was discussed earlier.

type can be either THRESH_BINARY, THRESH_BINARY_INV,
THRESH_TRUNC, THRESH_TOZERO, THRESH_TOZERO_INV.
Also, THRESH_OTSU can be added to any of the above.
For example, in THRESH_BINARY+THRESH_OTSU the
threshold value is determined by Otsu's method and
then that threshold value will be applied based on
the rules defined by THRESH_BINARY. The pixels with
intensities greater than the threshold value will
be assigned the maxval and the
rest will be assigned 0.

Returns: Output array same size and type as input array.

The cv2 function for distance transform is given below:

```
cv2.DistTransform(image, distance_Type, mask_Size,
labels, labelType)
```

Necessary arguments:

image is a 8-bit single channel image.

distance_Type is used to specify the distance formula.
It can be either CV_DIST_L1 (given by 0), CV_DIST_L2
(given by 1) or CV_DIST_C (given by 2). The distance
between (x,y) and (t,s) for CV_DIST_L1 is |x-t|+|y-s|
while CV_DIST_L2 is the Euclidean distance and
CV_DIST_C is the max{|x-t|,|y-s|}.

The size for the mask can be specified by mask_Size.
   If mask_Size is 3, a 3-by-3 mask is considered.

Optional arguments:

A 2D array of labels can be returned using labels.

The type of the above array of labels can be specified
by labelType. If labelType is DIST_LABEL_CCOMP,
then each connected component will be assigned the
same label. If labelType is DIST_LABEL_PIXEL then
each connected component will have its own label.

Returns: Output is a distance image same size as the input.

The cv2 function for watershed is given below:

```
cv2.watershed(image, markers)
```

Necessary arguments:
   image is the 8-bit 3 channel color image. Internally, the

function converts the color image to grayscale.
Only accepts color image as input.

markers is a labelled 32-bit single channel image.

Returns:
  Output is a 32 bit image. Output is overwritten on the
marker image.

   The cv2 code for the watershed segmentation is given below. The
various Python statements leading to the call to the cv2.watershed function create the marker image. The image in Figure 8.7(a)) shows dyed
osteoblast cells cultured in a bottle. The image is read and thresholded
(Figure 8.7(b)) to obtain foreground pixels. The image is converted
to a grayscale image before thresholding. The image is eroded (Figure 8.7(c)) to ensure that guaranteed foreground pixels are obtained.
Distance transform (Figure 8.7(d)) and the corresponding thresholding (Figure 8.7(e)) ensures the guaranteed foreground pixel image (i.e.,
marker image) is obtained. The marker image is used in the watershed
to obtain the image shown in Figure 8.7(f). The inputs for the cv2
watershed function are input image as a color image and the marker
image.

```
import cv2
from scipy.ndimage import label

# Opening the image.
a = cv2.imread('../Figures/cellimage.png')
# Converting to grayscale.
a1 = cv2.cvtColor(a, cv2.COLOR_BGR2GRAY)
# Thresholding the image to obtain cell pixels.
thresh,b1 = cv2.threshold(a1, 0, 255,
            cv2.THRESH_BINARY_INV+cv2.THRESH_OTSU)
```

```
# Since Otsu's method has over segmented the image
# erosion operation is performed.
b2 = cv2.erode(b1, None,iterations = 2)
# Distance transform is performed
dist_trans = cv2.distanceTransform(b2, 2, 3)
# Thresholding the distance transform image to obtain
# pixels that are foreground.
thresh, dt = cv2.threshold(dist_trans, 1,
            255, cv2.THRESH_BINARY)
# Performing labeling.
labelled, ncc = label(dt)
# Performing watershed.
cv2.watershed(a, labelled)
# Saving the image as watershed_output.png
cv2.imwrite('../Figures/watershed_output.png', labelled)
```

8.4   Contour-Based Segmentation

8.4.1   Chan-Vese Segmentation

Chan-Vese ([CV99]) is a region segmentation technique. It poses segmentation as an optimization problem. It allows segmentation even if the boundary between objects is not well defined.

Consider an image $f(x)$ where x could have multiple dimensions. Also assume there exists a curve (C) on this image. We will find the best fitting curve by minimizing Equation 8.5. The first term ensures that the curve has the minimum length. The second term ensures that the curve has the smallest area. The third term is evaluated only inside the curve and it ensures that all pixels inside the curve have a pixel value close to c_1. The fourth term is evaluated only on the outside of

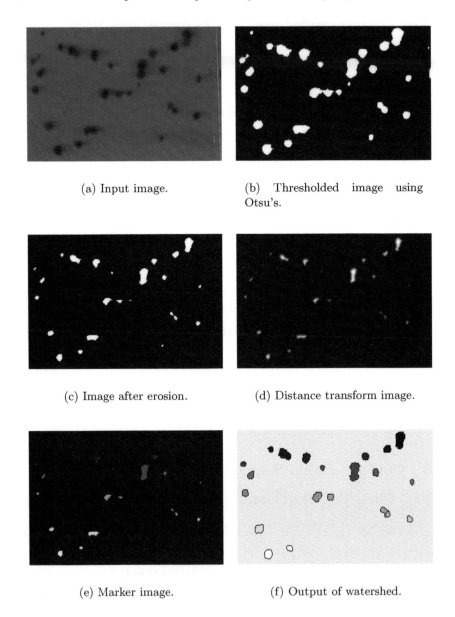

(a) Input image.

(b) Thresholded image using Otsu's.

(c) Image after erosion.

(d) Distance transform image.

(e) Marker image.

(f) Output of watershed.

FIGURE 8.7: An example of watershed segmentation. Original image reprinted with permission from Dr. Susanta Hui, Masonic Cancer Center, University of Minnesota.

the curve and it ensures that all pixels outside the curve have a pixel value close to c_2.

$$\underset{c_1,c_2,C}{\operatorname{argmin}} \mu \, Length(C) + \nu \, InsideArea(C) + \lambda_1 \int |f(x) - c_1| + \lambda_2 \int |f(x) - c_2| \tag{8.5}$$

The term μ determines the smoothness of the curve. A higher value produces a smoother curve while a smaller value close to zero produces a rough curve. In the creation of a smoother curve, smaller regions are excluded while the rough curve includes smaller objects.

Typically, λ_1 and λ_2 are equal in weighting the region inside and outside equally. The default value for both is 1 in scikit-image.

The implementation of Chan-Vese in scikit-image does not include the second term in Equation 8.5.

The following program demonstrates segmentation using the Chan-Vese algorithm. The image is read and converted to gray-scale, as scikit-image can only perform this segmentation on gray-scale images. The image is supplied to the chan_vese function with 3 possible values (0.1, 0.3, 0.6) for μ. The rest of the code plots the input image and the output image corresponding to the various μ.

```
from PIL import Image
import matplotlib.pyplot as plt
from skimage.segmentation import chan_vese
import numpy as np


# Opening the image and converting it into grayscale
img = Image.open('../Figures/imageinverse_input.png').
        convert('L')
img = np.array(img)


cv1 = chan_vese(img, mu=0.1)
cv2 = chan_vese(img, mu=0.3)
cv3 = chan_vese(img, mu=0.6)
```

```
fig, axes = plt.subplots(2, 2, figsize=(8, 8))
ax = axes.flatten()
ax[0].imshow(img, cmap="gray")
ax[0].set_axis_off()
ax[0].set_title("Original Image", fontsize=12)


ax[1].imshow(cv1, cmap="gray")
ax[1].set_axis_off()
ax[1].set_title("mu=0.1", fontsize=12)


ax[2].imshow(cv2, cmap="gray")
ax[2].set_axis_off()
ax[2].set_title("mu=0.3", fontsize=12)



ax[3].imshow(cv3, cmap="gray")
ax[3].set_axis_off()
ax[3].set_title("mu=0.6", fontsize=12)
plt.show()
```

The output of the program is in Figure 8.8. The top left image is the original image. The top right is the segmentation with $\mu = 0.1$. The bottom left is the segmentation with $\mu = 0.3$ and the bottom right is the segmentation with $\mu = 0.6$.

As discussed earlier, a smaller value of μ (such as 0.1) produces a smoother curve and only finds a larger object in the segmentation. The larger values of μ (such as 0.6) produce a rough curve and find smaller objects as well.

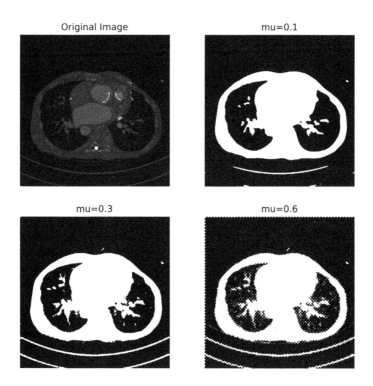

FIGURE 8.8: Chan-Vese segmentation and the effect of μ.

8.5 Segmentation Algorithm for Various Modalities

So far, we have discussed a few segmentation algorithms without concerning ourselves with the imaging modalities. Each imaging modality has unique characteristics that need to be understood in order to create a good segmentation algorithm.

8.5.1   Segmentation of Computed Tomography Image

The details of CT imaging are discussed in Chapter 13, "X-Ray and Computer Tomography." In a CT image, the pixel intensities are in Hounsfield units. The pixel intensities have physical significance as they are a map of the electron density of that material. The units are the same whether we image a human being, a mouse or a dog. Thus, a pixel intensity of +1000 always corresponds to a material that has electron density similar to bone. A pixel intensity of −1000 always corresponds to a material that has electron density similar to air. Hence, the segmentation process becomes simpler in the case of CT. To segment bone in a CT image, a simple thresholding such as assigning all pixels with values greater than +1000 being assigned 1 will suffice. A list of the range of pixel values corresponding to various materials such as soft tissue, hard tissue, etc., are available and hence simplify the segmentation process. This, however, assumes that the CT image has been calibrated to a Hounsfield unit. If not, traditional segmentation techniques have to be used.

8.5.2   Segmentation of MRI Image

The details of MRI are discussed in Chapter 14, "Magentic Resonance Imaging." MRI images do not have a standardized unit and hence need to be segmented using more traditional segmentation techniques discussed in this chapter.

8.5.3   Segmentation of Optical and Electron Microscope Images

The details of optical and electron microscope are discussed in Chapter 15, "Light Microscopes" and Chapter 16, "Electron Microscopes," respectively. In CT and MRI imaging of patients, the shape, size and position of organs remain similar across patients. In the case of optical and electron microscopes, two images acquired from the same specimen may not look alike and hence traditional techniques have to be used.

8.6 Summary

- Segmentation is a process of separating an image into multiple logical segments.

- Histogram-based method determines the threshold based on a histogram.

- Otsu's method determines the threshold that maximizes the variance between groups or minimizes the variance within a group.

- The threshold that maximizes the entropy between the foreground and background is the Renyi entropy threshold.

- The adaptive thresholding method segments the image by dividing the image into sub-images and then applying thresholding to each sub-image.

- Watershed segmentation is used when there are overlapping objects in an image.

- Chan-Vese segmentation is an optimization problem that draws a curve with the smallest length and area.

8.7 Exercises

1. In this chapter, we discussed a few segmentation methods. Consult the books listed as references and explain at least three more methods including details of the segmentation process, its advantages and disadvantages.

2. Consider any of the images used in histogram-based segmentation in this chapter. Rotate or translate the image using ImageJ by various angles and distance, and for each case segment the image. Are the threshold values different for different levels of rotation and translation? If there are differences in threshold value, explain the cause of the changes.

3. What happens if you zoom into the image using ImageJ while keeping the image size the same? Try different zoom levels (2X, 3X, and 4X). Explain the cause of change in threshold value.

   Hint: This changes the content of the image significantly and hence the histogram and the segmentation threshold.

4. In the various segmentation results, you will find spurious objects. Suggest a method to remove these objects.

   Hint: Morphology.

Chapter 9

Morphological Operations

9.1 Introduction

So far, we have discussed the various methods for manipulating individual pixels in the image through filtering, Fourier transform, etc. An important part of image analysis involves understanding the shape of the objects in that image through morphological operations. Morphology means form or structure. In morphological operations, the goal is to transform the structure or form of the objects using a structuring element. These operations change the shape and size of the objects in the image. Morphological operations can be applied on binary, grayscale, and color images. We omit morphology on color images in this chapter, as most bio-medical images are grayscale or binary images. We begin with basic morphological operations such as dilation, erosion, opening, and closing and then progress to compound operations such as hit-or-miss and skeletonization.

9.2 History

Morphology was introduced by Jean Serra in the 1960s as a part of his Ph.D. thesis under Georges Matheron at the Ecole des Mines de Paris, France. Serra applied the techniques he developed in the field of geology. With the arrival of modern computers, morphology began to be applied on images of all types such as black and white, grayscale and color. Over the next several decades, Serra developed the formalism for applying morphology on various data types like images, videos, meshes, etc. More information can be found in [Dou92],[HBS13],[MB90],[NT10],[Ser82],[SS94],[Soi04].

9.3 Dilation

For this section, we will assume that there is a binary input image. The foreground pixels have a value of 1 while the background pixels have a value of 0. The dilation operation allows the foreground pixels in an image to grow or expand. Hence this operation will also fill small holes in an object. It is also used to combine objects that are close enough to each other but are not connected.

The dilation of the image I with a structuring element S is denoted as $I \oplus S$.

Figure 9.1(a) is a binary image of size 4-by-5. The foreground pixels have intensity of 1 while background pixels have intensity of 0. The structuring element, Figure 9.1(b), is used to perform the dilation. The dilation process is explained in detail in the following steps:

1. Figure 9.1(a) is the binary image with 0's and 1's as the input.

2. The structuring element that will be used for dilation is shown in Figure 9.1(b). The shaded square on the 1 represents the reference pixel or origin of the structuring element. In this case the structuring element is of size 1-by-2. Both values in the structuring element play an important role in the dilation process.

3. To better illustrate the dilation process, we consider the first row in Figure 9.1(a) and apply the structuring element on each pixel in that row.

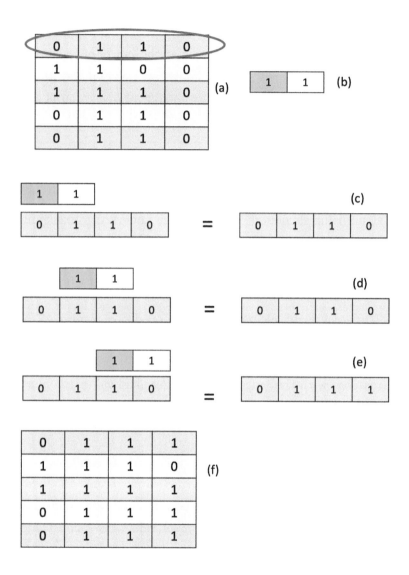

FIGURE 9.1: An example of binary dilation: (a) binary image for dilation, (b) structuring element, (c), (d), (e) application of dilation at various points in the image, and (f) final output after dilation.

4. With this structuring element, we can only grow the boundary by one more pixel to the right. If we considered a 1-by-3 structuring

element with all 1's and the origin of the structuring element at the center, then the boundary will grow by one pixel each in the left and right directions. Note that the morphological operations are performed on the input image and not on the intermediate results. The output of the morphological operation is the aggregate of all the intermediate results.

5. The structuring element is placed over the first pixel of the row and the pixel values in the structuring element are compared with the pixel values in the image. Since the reference value in the structuring element is 1, whereas the underlying pixel value in the image is 0, the pixel value in the output image remains unchanged. In Figure 9.1(c) the left side is the input to the dilation process and the right side is the intermediate result.

6. The structuring element is then moved one pixel over. Now the reference pixel in the structuring element and the image pixel value match. Since the value next to the reference value also matches with the 1 in the underlying pixel value, the pixel values in the output image do not change, and the output is shown in Figure 9.1(d).

7. The structuring element is then moved one pixel over. Now the reference pixel in the structuring element and the pixel value match. But the value next to the reference value does not match with the 0 in the underlying pixel value; the pixel value in the intermediate result will be changed to 1 as in Figure 9.1(e).

8. If the structuring element is then moved one pixel over, we will fall outside the image bounds.

9. This process is repeated on every pixel in the input image. The output of the dilation process on the whole image is given in Figure 9.1(f).

10. The process can be iterated multiple times using the same structuring element. In such case, the output from the previous iteration (Figure 9.1(f)) is used as input to the next iteration.

In summary, the dilation process first detects the boundary pixels of the object and it grows the boundary by a certain number of pixels (1 pixel to the right in this case). By repeating this process through multiple iterations or by using a large structuring element, the boundary pixels can grow by several pixels.

The following is the Python function for binary dilation:

```
scipy.ndimage.morphology.binary_dilation(input,
    structure=None,iterations=1,mask=None,
    output=None,border_value=0,
    origin=0,brute_force=False)

Necessary arguments:
  input = input image

Optional arguments:
   structure is the structuring element used for
the dilation, which was discussed earlier. If no
structure is provided, scipy assumes a square
structuring element of value 1.
The data type is ndarray.

   iterations are the number of times the dilation
operation is repeated. The default value is 1.
If the value is less than 1, the process is
repeated until there is no change in results.
The data type is integer or float.

   mask is an image, with the same size as the
input image with value of either 1 or 0.
```

Only points in the input image corresponding
to value of 1 in the mask image are modified
at each iteration. This is useful, if only a
portion of the input image needs to be dilated.
The data type is an ndarray.

  origin determines origin of the structuring
element, structure. The default value 0
corresponds to a structuring element whose
origin (reference pixel) is at the center.
The data needs to be either int for 1D
structuring element or tuples of int for
multiple dimension. Each value in the tuple
corresponds to different dimensions in the
structuring element.

  border_value will be used for the border pixels
in the output image. It can either be 0 or 1.

Returns: output as an ndarray.

The following is Python code that takes an input image and performs dilation with 5 iterations using the binary_dilation function:

```
from PIL import Image
import scipy.ndimage as snd
import numpy as np
import cv2


# Opening the image and converting it to grayscale.
a = Image.open('../figures/dil_image.png').convert('L')
a = np.array(a)
# Performing binary dilation for 5 iterations.
```

```
b = snd.morphology.binary_dilation(a, iterations=5)
# Saving the image as 8-bit as b is a
# binary image of dtype=bool
cv2.imwrite('../figures/di_binary.png', b*255)
```

Figure 9.2(a) is the input image for binary dilation with 5 iterations and the corresponding output image is given in Figure 9.2(b). Since binary dilation makes the foreground pixels dilate or grow, the small black spots (background pixels) inside the white regions (foreground pixels) in the input image disappear.

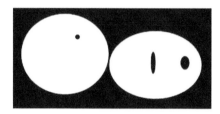

(a) Black and white image for dilation.

(b) Output image after dilation with 5 iterations.

FIGURE 9.2: An example of binary dilation.

9.4   Erosion

Erosion is used to shrink objects in an image by removing pixels from the boundary of that object. Erosion is opposite of dilation.

The erosion of the image I and with a structuring element S is denoted as $I \ominus S$.

Let us consider the same binary input and the structuring element that was considered for dilation to illustrate erosion. Figure 9.3(a) is a binary image of size 4 by 5. The structuring element 9.3(b) is used to

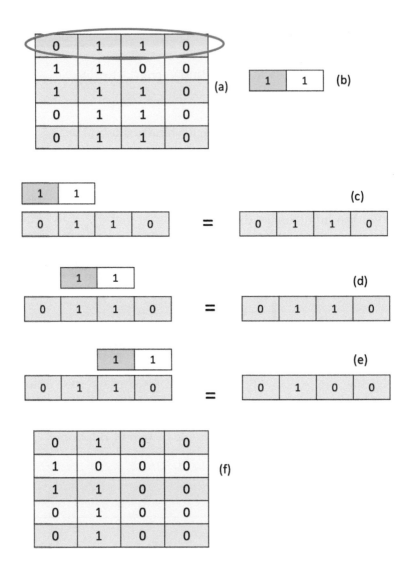

FIGURE 9.3: An example of binary erosion: (a) binary image for erosion, (b) structuring element, (c), (d), (e) application of erosion at various points in the image, and (f) final output after erosion.

perform the erosion. The erosion process is explained in detail in the following steps:

1. Figure 9.3(a) is an example of a binary image with 0's and 1's. The background pixels are represented by 0 and the foreground by 1.

2. The structuring element that will be used for erosion is shown in Figure 9.3(b). The shaded square on the 1 represents the reference pixel or origin of the structuring element. In this case the structuring element is of size 1-by-2. Both values in the structuring element play an important role in the erosion process.

3. Consider the first row in Figure 9.3(a) and apply the structuring element on each pixel of the row.

4. With this structuring element, we can only erode the boundary by one pixel to the right.

5. The structuring element is placed over the first pixel of that row and the pixel values in the structuring element are compared with the pixel values in the image. Since the reference value in the structuring element is 1, whereas the underlying pixel value in the image is 0, the pixel value remains unchanged. In Figure 9.3(c), the left side is the input to the erosion process and the right side is the intermediate output.

6. The structuring element is then moved one pixel over. The reference pixel in the structuring element and the image pixel value match. Since the value next to the reference value also matches with the 1 in the underlying pixel value, the pixel values in the output image do not change, as shown in Figure 9.3(d).

7. The structuring element is then moved one pixel over. The reference pixel in the structuring element and the image pixel value match, but the non-reference value does not match with the 0 in the underlying pixel value. The structuring element is on the boundary. Hence, the pixel value below the reference value is replaced with 0, as shown in Figure 9.3(e).

8. If the structuring element is then moved one pixel over, we will fall outside the image bounds.

9. This process is repeated on every pixel in the input image. The output of the erosion process on the whole image is given in Figure 9.3(f).

10. The process can be iterated multiple times using the same structuring element. In such case, the output from the previous iteration (Figure 9.3(f)) is used as input to the next iteration.

In summary, the erosion process first detects the boundary pixels of the object and shrinks the boundary by a certain number of pixels (1 pixel from the right in this case). By repeating this process through multiple iterations or by using a larger structuring element, the boundary pixels can be shrunk by several pixels.

The Python function for binary erosion is given below. The arguments for binary erosion are the same as the binary dilation arguments listed previously.

```
scipy.ndimage.morphology.binary_erosion(input,
  structure=None,iterations=1,mask=None,
  output=None,border_value=0,origin=0,
  brute_force=False)
```

The Python code for binary erosion is given below.

```
from PIL import Image
import scipy.ndimage as snd
import numpy as np
import cv2


# Opening the image and converting it to grayscale.
a = Image.open('../figures/er_image.png').convert('L')
```

```
a = np.array(a)
# Performing binary erosion for 20 iterations.
b = snd.morphology.binary_erosion(a,iterations=20)
# Saving the image as 8-bit as b is a
# binary image of dtype=bool
cv2.imwrite('../figures/er_binary_output_20.png', b*255)
```

Figures 9.4(b) and 9.4(c) demonstrate the binary erosion of 9.4(a) using 10 and 20 iterations respectively. Erosion removes boundary pixels and hence after 10 iterations, the two circles are separated creating a dumbbell shape. A more profound dumbbell shape is obtained after 20 iterations.

(a) Input image for erosion.

(b) Output image after 10 iterations.

(c) Output image after 20 iterations.

FIGURE 9.4: An example of binary erosion.

9.5 Grayscale Dilation and Erosion

Grayscale dilation and erosion are similar to their binary counterparts. In binary dilation and erosion, the foreground pixels in the input image have a pixel value of 1 while the background pixels have a pixel value of 0. In grayscale dilation and erosion, the foreground pixels and background pixels can take pixel values in the grayscale range. For example, we can supply an 8-bit image as an input.

In grayscale erosion, the bright pixel values will shrink and the dark pixels increase or grow. Small bright objects will be eliminated by grayscale erosion and dark objects will grow. The effect of erosion can be observed in region(s) where there is a change in the grayscale intensity.

The following is the Python function for grayscale erosion.

```
scipy.ndimage.morphology.grey_erosion(input,
footprint)
```

```
Necessary arguments:
input has to be an ndarray.
```

```
Optional arguments:
footprint is a structure element that is an ndarray of
integers.
```

```
Returns:
An ndarray.
```

The Python code for grayscale erosion is given below.

```
import numpy as np
from PIL import Image
import scipy.ndimage

# Opening the image and converting it into grayscale.
a = Image.open('../figures/sem3.png').convert('L')
# Creating a structuring element.
footprint = np.ones((15, 15))
# Performing grey erosion.
b = scipy.ndimage.morphology.grey_erosion(a,
    footprint=footprint)
```

```
# Converting ndarray to image.
c = Image.fromarray(b)
# Saving the image.
c.save('../figures/grey_erosion_output_15.png')
```

Figures 9.5(b) and 9.5(c) demonstrate the grayscale erosion of the image in Figure 9.5(a) using a 15-by-15 structuring element and a 25-by-25 structuring element respectively. Grayscale erosion increases the number of background pixels. In Figure 9.5(a), the foreground has bright regions, and there are small black holes, thin black lines, and dark holes. On the top right, the input image has a hole. After grayscale erosion with a 15-by-15 structuring element, notice that the hole on the top right, has shrunk. Also, the foreground pixels are no longer strongly connected. Finally, prominent black horizontal lines are introduced. Grayscale erosion with a 25-by-25 structuring element further shrinks the hole, the foreground pixels are further disconnected, and the black horizontal lines are prominent.

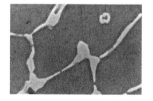 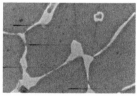 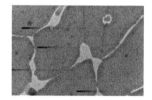

(a) Input image for grayscale erosion.

(b) Output image with a 15-by-15 structuring element.

(c) Output image with a 25-by-25 structuring element.

FIGURE 9.5: An example of grayscale erosion.

In grayscale dilation, bright pixels increase or grow and dark pixels decrease or shrink. The effect of dilation can be clearly observed in a region(s) where there is a change in the grayscale intensity. Similar to binary dilation, grayscale dilation fills holes.

The following is the Python function for grayscale dilation.

```
scipy.ndimage.morphology.grey_dilation(input,
footprint)
```

```
Necessary arguments:
input has to be an ndarray.
```

```
Optional arguments:
footprint is a structure element that is an ndarray of
integers.
```

```
Returns:
An ndarray.
```

The Python code for grayscale dilation is given below.

```python
import numpy as np
from PIL import Image
import scipy.ndimage

# Opening the image and converting it into grayscale.
a = Image.open('../figures/sem3.png').convert('L')
# Creating a structuring element.
footprint = np.ones((15,15))
# Performing grey dilation.
b = scipy.ndimage.morphology.grey_dilation(a,
    footprint=footprint)
# Converting ndarray to image.
c = Image.fromarray(b)
# Saving the image.
c.save('../figures/grey_dilation_output_15.png')
```

Figures 9.6(b) and 9.6(c) demonstrate the grayscale dilation of the image in Figure 9.6(a) using a 15-by-15 structuring element and a 25-by-25 structuring element respectively. Grayscale dilation increases the number of foreground pixels. In Figure 9.6(a), apart from the foreground pixels that are bright, there are small black holes, thin black lines and dark holes. On the top right, there is a hole. After grayscale dilation with a 15-by-15 structuring element, notice that the hole in the top right grew a bit, the foreground pixels became thick and some black lines were removed. After grayscale dilation with a 25-by-25 structuring element, the hole in the top right grew bigger, the foreground pixels are thicker and the lines disappear.

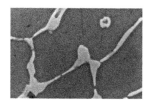 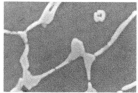 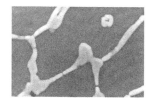

(a) Input image for grayscale dilation.

(b) Output image with a 15-by-15 structuring element.

(c) Output image with a 25-by-25 structuring element.

FIGURE 9.6: An example of grayscale dilation.

9.6 Opening and Closing

Opening and closing operations are complex morphological operations. They are obtained by combining dilation and erosion. Opening and closing can be performed on binary, grayscale and color images.

Opening is defined as erosion followed by dilation of an image. The opening of the image I with a structuring element S is denoted as

$$I \circ S = (I \ominus S) \oplus S \tag{9.1}$$

Closing is defined as dilation followed by erosion of an image. The closing of the image I with a structuring element S is denoted as

$$I \bullet S = (I \oplus S) \ominus S \qquad (9.2)$$

The following is the Python function for opening:

```
scipy.ndimage.morphology.binary_opening(input,
structure=None, iterations=1, output=None, origin=0)


Necessary arguments:
input = array

Optional arguments:
    structure is the structuring element used for
    the dilation, which was discussed earlier. If no
    structure is provided, scipy assumes a square
    structuring element of value 1.
    The data type is ndarray.

    iterations are the number of times the opening is
    performed (erosion followed by dilation). The
    default value is 1. If the value is less than 1,
    the process is repeated until there is no change
    in results. The data type is integer or float.

    origin determines origin of the structuring element.
    The default value 0 corresponds to a
    structuring element whose origin (reference pixel)
    is at the center. The data needs to be either int
    for 1D structuring element or tuples of int for
    multiple dimension. Each value in the tuple
```

corresponds to different dimensions in the
structuring element.

Returns: output as an ndarray

The Python code for binary opening with 5 iterations is given below.

```
from PIL import Image
import scipy.ndimage as snd
import numpy as np
import cv2

# Opening the image and converting it to .
a = Image.open('../figures/dil_image.png').convert('L')
a = np.array(a)
# Defining the structuring element.
s = [[0,1,0],[1,1,1], [0,1,0]]
# Performing the binary opening for 5 iterations.
b = snd.morphology.binary_opening(a, structure=s,
    iterations=5)
# Saving the image as 8-bit as b is a
# binary image of dtype=bool
cv2.imwrite('../figures/opening_binary.png', b*255)
```

Figure 9.7(b) is the output of the binary opening with 5 iterations. Binary opening has altered the boundaries of the foreground objects. The size of the small black holes inside the objects has also changed.

The Python function for binary closing is given below. The arguments for binary closing are the same as the binary opening arguments.

```
scipy.ndimage.morphology.binary_closing(input,
 structure=None, iterations=1,output=None, origin=0)
```

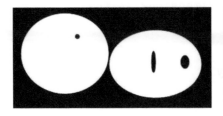 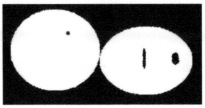

(a) Input image for opening.   (b) Output image after opening.

FIGURE 9.7: An example of binary opening with 5 iterations.

The Python code for closing is given below and an example is given in Figure 9.8. The closing operation has resulted in filling in the holes, as shown in Figure 9.8(b).

```
from PIL import Image
import scipy.ndimage as snd
import numpy as np
import cv2

# Opening the image and converting it to grayscale.
a = Image.open('../figures/dil_image.png').convert('L')
a = np.array(a)
# Defining the structuring element.
s = [[0,1,0],[1,1,1], [0,1,0]]
# Performing the binary closing for 5 iterations.
b = snd.morphology.binary_closing(a,structure=s,
    iterations=5)
# Saving the image as 8-bit as b is a
# binary image of dtype=bool
cv2.imwrite('../figures/closing_binary.png', b*255)
```

It can be observed that the black holes in the input image are elongated after the opening operation, while the closing operation on the same input filled the holes.

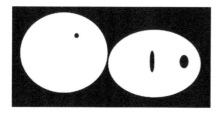 

(a) Input image for closing.   (b) Output image after closing.

FIGURE 9.8: An example of binary closing with 5 iterations.

9.7 Grayscale Opening and Closing

Grayscale opening and closing are similar to their binary counterparts.

Erosion followed by dilation results in opening.

The following is the Python function for grayscale opening.

```
scipy.ndimage.morphology.grey_opening(input,
footprint)
```

```
Necessary arguments:
input has to be an ndarray.
```

```
Optional arguments:
footprint is a structure element that is an ndarray of
integers.
```

```
Returns:
An ndarray.
```

The Python code for grayscale erosion is given below.

```
import numpy as np
from PIL import Image
import scipy.ndimage


# Opening the image and converting it into grayscale.
a = Image.open('../figures/adaptive_example1.png').
convert('L')
# Creating a structuring element.
footprint = np.ones((40,40))
# Performing grey opening.
b = scipy.ndimage.morphology.grey_opening(a,
    footprint=footprint)
# Converting ndarray to image.
c = Image.fromarray(b)
# Saving the image.
c.save('../figures/grey_opening_output_40.png')
```

Figure 9.9(a) is the input image that we will use for grayscale opening. After performing grayscale opening with a 40-by-40 structuring element, we obtain Figure 9.9(b). Notice that the opening operation was able to detect the regions where there was text in the input image.

(a) Input image for grayscale opening.

(b) Output image with a 40-by-40 structuring element.

FIGURE 9.9: An example of grayscale opening.

Dilation followed by erosion results in closing.

The following is the Python function for grayscale closing.

```
scipy.ndimage.morphology.grey_closing(input,
footprint)


Necessary arguments:
input has to be an ndarray.


Optional arguments:
footprint is a structure element that is an ndarray of
integers.


Returns:
An ndarray.
```

The Python code for grayscale closing is given below.

```
import numpy as np
from PIL import Image
import scipy.ndimage


# Opening the image and converting it into grayscale.
a = Image.open('../figures/adaptive_example1.png').
convert('L')
a = np.asarray(a)
# Creating a structuring element.
fp = np.ones((40,40))
# Performing grey closing.
bg = scipy.ndimage.morphology.grey_closing(a,
    footprint=fp)
# bg represents the background.
```

```
# We will subtract bg from a to remove the background in a.
bg_free = (a.astype(np.float64) - bg.astype(np.float64))
# We rescale bg_free to 0 to 255.
denom = (bg_free.max()-bg_free.min())
bg_free_norm = (bg_free - bg_free.min())*255/denom
# Converting bg_free_norm to uint8.
bg_free_norm = bg_free_norm.astype(np.uint8)
# Converting bg_free_norm and bg to images.
bg_free_norm = Image.fromarray(bg_free_norm)
bg = Image.fromarray(bg)
# Saving the background image.
bg.save('../figures/grey_closing_out_40.png')
# Saving the bg_free_norm image.
bg_free_norm.save('../figures/closing_bgfree.png')
```

Figure 9.10(a) is the input image that we consider for grayscale closing. Figure 9.10(b) was obtained after applying grayscale closing with a 40-by-40 structuring element. This represents the background of the original image. When we subtract the background image from the original image, then we obtain Figure 9.10(c). Notice that after subtraction, the background is uniform. We have hence achieved background subtraction using grayscale closing.

(a) Input image for grayscale closing.

(b) Output image with a 40-by-40 structuring element.

(c) The difference between the left image and the middle image.

FIGURE 9.10: An example of grayscale closing.

9.8 Hit-or-Miss

Hit-or-miss transformation is a morphological operation used in finding specific patterns in an image. Hit-or-miss is used to find boundary or corner pixels, and is also used for thinning and thickening, which are discussed in the next section. Unlike the methods we have discussed so far, this method uses more than one structuring element and all its variations to determine pixels that satisfy a specific pattern.

Let us consider a 3-by-3 structuring element with origin at the center. The structuring element with 0's and 1's shown in Table 9.1 is used in the hit-or-miss transformation to determine the corner pixels. The blank space in the structuring element can be filled with either 1 or 0.

TABLE 9.1: Hit-or-miss structuring element

	1	
0	1	1
0	0	

Since we are interested in finding the corner pixels, we have to consider all four variations of the structuring element in Table 9.1. The four structuring elements given in Table 9.2 will be used in the hit-or-miss transformation to find the corner pixels. The origin of the structuring element is applied to all pixels in the image and the underlying pixel values are compared. As discussed in Chapter 4 on filtering, the structuring element cannot be applied to the edges of the image. So the edges of the image are assumed to be zero in the output.

After determining the locations of the corner pixels from each structuring element, the final output of hit-or-miss is obtained by performing an OR operation on all the output images.

Let us consider a binary image in Figure 9.11(a). After performing the hit-or-miss transformation on this image with the structuring elements in Table 9.2, we obtain the image in Figure 9.11(b). Notice

TABLE 9.2: Variation of all structuring elements used to find corners.

	1	
0	1	1
	0	

	1	
1	1	0
	0	0

0	0	
0	1	1
	1	

	0	0
1	1	0
	1	

0	0	0	0	0	0	0	0	0	0	0
0	0	0	1	0	1	0	0	0	0	0
0	0	1	1	0	1	0	0	0	0	0
0	0	1	1	1	1	1	0	0	0	0
0	0	1	1	1	1	1	0	0	0	0
0	0	1	0	0	0	1	1	0	0	0
0	0	1	0	0	0	1	1	0	0	0
0	0	1	1	1	1	1	0	0	0	0
0	0	1	1	1	1	0	0	0	0	0
0	0	0	0	0	0	0	0	0	0	0
0	0	0	0	0	0	0	0	0	0	0

0	0	0	0	0	0	0	0	0	0	0
0	0	0	0	0	0	0	0	0	0	0
0	0	1	0	0	0	0	0	0	0	0
0	0	0	0	0	0	1	0	0	0	0
0	0	0	0	0	0	1	0	0	0	0
0	0	0	0	0	0	0	1	0	0	0
0	0	0	0	0	0	0	1	0	0	0
0	0	0	0	0	0	1	0	0	0	0
0	0	1	0	0	0	0	0	0	0	0
0	0	0	0	0	0	0	0	0	0	0
0	0	0	0	0	0	0	0	0	0	0

(a) Input image for hit-or-miss.

(b) Output image of hit-or-miss.

FIGURE 9.11: An example of hit-or-miss transformation.

that the pixels in the output of Figure 9.11(b) are a subset of boundary pixels.

The following is the Python function for hit-or-miss transformation:

```
scipy.ndimage.morphology.binary_hit_or_miss(input,
   structure1=None, structure2=None,
output=None, origin1=0, origin2=None)

Necessary arguments:
input is a binary array

Optional arguments:
   structure1 is a structuring element that is used
   to fit the foreground of the image.  If no
   structuring element is provided, then scipy will
   assume square structuring element of value 1.
```

structure2 is a structuring element that is used to miss the foreground of the image. If no structuring element is provided, then scipy will consider a complement of structuring element provided in structure1.

origin1 determines origin of the structuring element, structure1. The default value 0 corresponds to a structuring element whose origin (reference pixel) is at the center. The data needs to be either int for 1D structuring element or tuples of int for multiple dimension. Each value in the tuple corresponds to different dimensions in the structuring element.

origin2 determines origin of the structuring element, structure2. The default value 0 corresponds to a structuring element whose origin (reference pixel) is at the center. The data needs to be either int for 1D structuring element or tuples of int for multiple dimension. Each value in the tuple corresponds to different dimensions in the structuring element.

Returns: output as an ndarray.

The Python code for hit-or-miss transform is given below.

```
from PIL import Image
import numpy as np
import scipy.ndimage as snd
import cv2
```

```
# Opening the image and converting it to grayscale.
a = Image.open('../figures/thickening_input.png').
convert('L')
a = np.array(a)


# Defining the structuring element.
structure1 = np.array([[1, 1, 0], [1, 1, 1],
            [1, 1, 1]])
# Performing the binary hit-or-miss.
b = snd.morphology.binary_hit_or_miss(a,
    structure1=structure1)
# Saving the image as 8-bit as b is a
# binary image of dtype=bool
cv2.imwrite('../figures/hitormiss_output2.png', b*255)
```

In the above program, a structuring element 'structure1' is created with all the elements listed and used in the hit-or-miss transformation. Figure 9.12(a) is the input image for the hit-or-miss transform and the corresponding output is in Figure 9.12(b). Notice that only a few boundary pixels from each object in the input image are identified by the hit-or-miss transformation. It is important to make a judicious choice of the structuring element in the hit-or-miss transform, as different elements have different effect on the output.

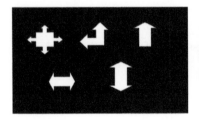 

(a) Input image for hit-or-miss   (b) Output image of hit-or-miss

FIGURE 9.12: An example of hit-or-miss transformation on a binary image.

9.9 Thickening and Thinning

Thickening and thinning transformations are an extension of hit-or-miss transformation and can only be applied to binary images.

Thickening is used to grow the foreground pixels in a binary image and is similar to the dilation operation. In this operation, the background pixels are added to the foreground pixels to make the selected region grow or expand or thicken. The thickening operation can be expressed in terms of the hit-or-miss operation. Thickening of the image I with the structuring element S can be given by Equation 9.3 where H is the hit-or-miss on image I with S,

$$\text{Thickening(I)} = I \cup H \qquad (9.3)$$

In the thickening operation, the origin of the structuring element has to be either zero or empty. The origin of the structuring element is applied to every pixel in the image (except the edges of the images). The pixel values in the structuring element are compared to the underlying pixels in the sub-image. If all the values in the structuring element match the pixel values in the sub-image, then the underlying pixel below the origin is set to 1 (foreground). In all other cases, it remains unchanged. In short, the output of the thickening operation consists of the original image and the foreground pixels that have been identified by the hit-or-miss transformation.

Thinning is the opposite of thickening. Thinning is used to remove selected foreground pixels from the image. Thinning is similar to erosion or opening as the thinning operation will result in the shrinking of foreground pixels. The thinning operation can also be expressed in terms of hit-or-miss transformation. The thinning of image I with the structuring element S can be given by Equation 9.4 where H is the hit-or-miss of image I with S,

$$\text{Thinning(I)} = I - H \qquad (9.4)$$

In the thinning operation, the origin of the structuring element has to be either 1 or empty. The origin of the structuring element is applied to every pixel in the image (except the edges of the images). The pixel values in the structuring element are compared to the underlying pixels in the image. If all the values in the structuring element match with the pixel values in the image, then the underlying pixel below the origin is set to 0 (background). In all other cases, it remains unchanged.

Both thickening and thinning operations can be applied repeatedly.

9.9.1 Skeletonization

The process of applying the thinning operation multiple times so that only connected pixels are retained is known as skeletonization. This is a form of erosion where most of the foreground pixels are removed and only pixels with connectivity are retained. As the name suggests, this method can be used to define the skeleton of the object in an image.

The following is the Python function for skeletonization:

```
skimage.morphology.skeletonize(image)
```

```
Necessary arguments:
image can be ndarray array of either binary or
boolean type. If the image is binary, foreground
pixels are represented by 1 and background pixels
by 0. If the image is boolean, True represents
foreground while false represents background.
```

```
Returns: output as an ndarray containing the skeleton
```

The Python code for skeletonization is given below.

```
import numpy as np
from PIL import Image
```

```
from skimage.morphology import skeletonize
import cv2

# Opening the image and converting it to grayscale.
a = Image.open('../figures//steps1.png').convert('L')
# Converting a to an ndarray and normalizing it.
a = np.asarray(a)/np.max(a)
# Performing skeletonization.
b = skeletonize(a)
# Saving the image as 8-bit as b is a
# binary image of dtype=bool
cv2.imwrite('../figures/skeleton_output.png', b*255)
```

Figure 9.13(a) is the input image for the skeletonization and Figure 9.13(b) is the output image. Notice that the foreground pixels have shrunk and only the pixels that have connectivity survive the skeletonization process. One of the major uses of skeletonization is in measuring the length of objects. Once the foreground pixels have been shrunk to one pixel width, the length of the object is approximately the number of pixels after skeletonization.

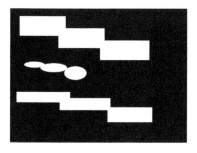

(a) Input image for skeletonization.

(b) Output image after skeletonization.

FIGURE 9.13: An example of skeletonization.

9.10   Summary

- The structuring element is important for most of the binary operations.

- Binary or grayscale dilation, closing, and thickening operations increase the number of foreground pixels and hence close holes in objects and aggregate nearby objects. The exact effect depends on the structuring element. The closing operation may preserve the size of the object while dilation does not.

- Binary or grayscale erosion, opening, and thinning operations decrease the number of foreground pixels and hence increase the size of holes in objects and also separate nearby objects. The exact effect depends on the structuring element. The opening operation may preserve the size of the object while erosion does not.

- Hit-or-miss transformation is used to determine specific patterns in an image.

- Skeletonization is a type of thinning operation in which only connected pixels are retained.

9.11   Exercises

1. Perform skeletonization on the image in Figure 9.2(a).

2. Consider an image and prove that erosion followed by dilation is not same as dilation followed by erosion.

3. Imagine an image containing two cells that are next to each other with a few pixels overlapping; what morphological operation would you use to separate them?

4. You are hired as an image processing consultant to design a new checkout machine. You need to determine the length of each vegetable programmatically given an image containing one of the vegetables. Assuming that the vegetables are placed one after the other, what morphological operation will you need?

Chapter 10

Image Measurements

10.1 Introduction

So far we have shown methods to segment an image and obtain various regions that share similar characteristics. The next step is to understand the shape, size and geometrical characteristics of these regions.

The regions in an image may be circular, such as an image of coins, or edges in a building. In some cases, the regions may not have simple geometrical shapes like circles, lines, etc. Hence radius, slope, etc., alone do not suffice to characterize the regions. An array of properties such as area, bounding box, central moments, centroid, eccentricity, Euler number, etc., are needed to describe shapes of regions.

In this chapter we begin the discussion with the label function that allows numbering each region uniquely, so that the 'regionprops' function can be used to obtain the characteristics. This is followed by the Hough transform for characterizing lines and circles. We will discuss a method for counting regions or objects using template matching. We conclude with a discussion of FAST and Harris corner.

10.2   Labeling

Labeling is used to identify different objects in an image. The image has to be segmented before labeling can be performed. In a labeled image, all pixels in a given object have the same value. For example, if an image comprises four objects, then in the labeled image, all pixels in the first object have a value 1, etc.

The Python function for labeling is given below.

```
skimage.morphology.label(image)
```

```
Necessary arguments:
   image is the segmented image as an ndarray.
```

```
Returns: output labelled image as an ndarray.
```

The Python function for obtaining geometrical characteristics of regions is regionprops. A labeled image is used as an input image to this function. Some of the parameters for regionprops are listed below. The complete list can be found at [Si20].

```
skimage.measure.regionprops
   (label_image)
Necessary arguments:
   label_image is a labelled image as an ndarray.
```

```
Returns:
A list of RegionProperties.
Let rprops be a list of region properties.
The rprops[0].area will return the
```

```
area of rprops[0], the first region.
The rprops[0].bbox will return the bounding
box of rprops[0].
```

The following is the Python code for obtaining the properties of various regions using regionprops. The input image is read and thresholded using Otsu's method. The various objects are labeled using the label function. At the end of this process, all pixels in a given object have the same pixel value. The labeled image is then given as an input to the regionprops function. The regionprops function calculates the area, centroid and bounding box for each of these regions. Finally, a loop is used to iterate through every region in regionprops output. For each region, the centroid and bounding box are marked on the image using matplotlib functions.

```python
import numpy
import cv2
from PIL import Image
import matplotlib.pyplot as plt
import matplotlib.patches as mpatches
from skimage.morphology  import label
from skimage.measure import regionprops
from skimage.filters.thresholding import threshold_otsu


# Opening the image and converting it to grayscale.
a = Image.open('../Figures/objects.png').convert('L')
# a is converted to an ndarray.
a = numpy.asarray(a)
# Threshold value is determined by
# using Otsu's method.
thresh = threshold_otsu(a)
# The pixels with intensity greater than
# "theshold" are kept.
```

```python
b = a > thresh
# Labelling is performed on b.
c = label(b)
# c is saved as label_output.png
cv2.imwrite('../Figures/label_output.png', c)
# On the labelled image c, regionprops is performed
d = regionprops(c)
# the following command creates an empty plot of
# dimension 6 inch by 6 inch
fig, ax = plt.subplots(ncols=1,nrows=1,
          figsize=(6, 6))
# plots the label image on the
# previous plot using colormap
ax.imshow(c, cmap='YlOrRd')


for i in d:
    # Printing the x and y values of the
    # centroid where centroid[1] is the x value
    # and centroid[0] is the y value.
    print(i.centroid[1], i.centroid[0])
    # Plot a red circle at the centroid, ro stands
    # for red.
    plt.plot(i.centroid[1],i.centroid[0],'ro')
    # In the bounding box, (lr,lc) are the
    # co-ordinates of the lower left corner and
    # (ur,uc) are the co-ordinates
    # of the top right corner.
    lr, lc, ur, uc = i.bbox
    # The width and the height of the bounding box
    # is computed.
    rec_width = uc - lc
    rec_height = ur - lr
```

```
    # Rectangular boxes with
# origin at (lr,lc) are drawn.
    rect = mpatches.Rectangle((lc, lr),rec_width,
            rec_height,fill=False,edgecolor='black',
            linewidth=2)
    # This adds the rectangular boxes to the plot.
    ax.add_patch(rect)
# Saving the figure
plt.savefig('../Figures/regionprops_output.png')
plt.show()
```

Figure 10.1(a) is the input image for the regionprops and Figure 10.1(b) is the output image. The output image is labeled with different colors and enclosed in a bounding box obtained using regionprops.

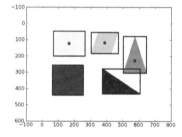

(a) Input image for region-props.

(b) Labeled output image with bounding boxes and centorids.

FIGURE 10.1: An example of regionprops.

10.3 Hough Transform

The edge-detection process discussed in Chapter 4, "Spatial Filters," detects edges in an image but does not characterize the slope and intercept of the line or the radius of a circle. These characteristics can be calculated using the Hough transform.

10.3.1   Hough Line

The general form of a line is given by $y = mx + b$, where m is the slope of the line and b is the y-intercept. But in the case of a vertical line, m is undefined or infinity and hence the accumulator plane (discussed below) will have infinite length, which cannot be programmed in a computer. Hence, we use polar coordinates which are finite for all slopes and intercepts to characterize a line.

The polar form of a line (also called the normal form) is given by the following equation:

$$x \cos(\theta) + y \sin(\theta) = r \tag{10.1}$$

where r is positive and is the perpendicular distance between the origin and the line and θ is the slope of the line and it ranges from $[0, 180]$. Each point in the (x, y) plane, also known as the Cartesian plane, can be transformed into the (r, θ) plane, also known as the accumulator plane, which is a 2D matrix with two coordinates r and θ.

A segmented image is given as an input for the Hough line transform. To characterize the line, a 2D accumulator plane with r and θ is generated. For a specific (r, θ) and for each x value in the image, the corresponding y value is computed using Equation 10.1. For every y value that is the foreground pixel i.e., the y value lies on the line, a value of 1 is added to the specific (r, θ) in the accumulator plane. This process is repeated for all values of (r, θ). The resultant accumulator plane will have high intensity at the points corresponding to a line. Then the (r, θ) corresponding to the local peak will provide the parameters of the line in the original image.

If the input image is of size N-by-N, the number of values of r is M and number of points in θ is K, the computational time for accumulator array is $O(KMN^2)$. Hence, the Hough line transform is a computationally intensive process. If θ ranges from $[0, 180]$ and for a step size of 1, then $K = 180$ along the θ axis. If the range of θ is known a priori and is smaller than $[0, 180]$, K will be smaller and hence the computation can be made faster. Similarly, if other factors such as M or N can be reduced, the computational time can be reduced as well.

The cv2 function for the Hough line transform is given below:

```
cv2.HoughLines(image,rho,theta,threshold)

Necessary argument:
image should be binary.

rho is the resolution of the distance in pixels
in the accumulator matrix.

theta is the resolution of the angle in pixels.

threshold is the minimum value that will be used
to detect a line in the accumulator matrix.

Returns: Outputs is a vector with distance and
   angle of detected lines.
```

The cv2 code for the Hough line transform is given below. The input image (Figure 10.2(a)) is converted to grayscale. The image is then thresholded using Otsu's method (Figure 10.2(b)) to obtain a binary image. On the thresholded image, Hough line transformation is performed. The output of the Hough line transform with the detected lines is shown in Figure 10.2(c). The thick lines are lines that are detected by the Hough line transform.

```
import cv2
import numpy as np
```

```
# Opening the image.
im = cv2.imread('../Figures/hlines.png')
# Converting the image to grayscale.
a1 = cv2.cvtColor(im, cv2.COLOR_BGR2GRAY)
# Thresholding the image to obtain
# only foreground pixels.
thresh, b1 = cv2.threshold(a1, 0, 255,
             cv2.THRESH_BINARY_INV+cv2.THRESH_OTSU)

cv2.imwrite('../Figures/hlines_thresh.png', b1)
# Performing the Hough lines transform.
lines = cv2.HoughLines(b1, 10, np.pi/20, 200)
for rho, theta in lines[0]:
    a = np.cos(theta)
    b = np.sin(theta)
    x0 = a*rho
    y0 = b*rho
    x1 = int(x0 + 1000*(-b))
    y1 = int(y0 + 1000*(a))
    x2 = int(x0 - 1000*(-b))
    y2 = int(y0 - 1000*(a))

    cv2.line(im,(x1,y1),(x2,y2),(0,0,255),2)

cv2.imwrite('../Figures/houghlines_output.png', im)

# Printing the lines: distance and angle in radians.
print(lines)
```

10.3.2   Hough Circle

The general form of a circle is given by $(x - a)^2 + (y - b)^2 = R^2$ where (a, b) is the center of the circle and R is the radius of the circle.

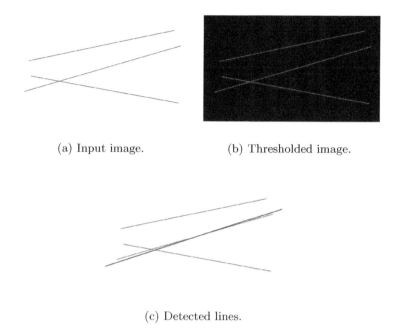

(a) Input image.   (b) Thresholded image.

(c) Detected lines.

FIGURE 10.2: An example of the Hough line transform.

The equation can be rewritten as $y = b \pm \sqrt{R^2 - (x-a)^2}$. Alternately, it can be written in polar form as

$$\begin{cases} x = a + R\cos(\theta) \\ y = b + R\sin(\theta) \end{cases} \tag{10.2}$$

where θ ranges from $[0, 360]$.

It can be seen from Equation 10.2 that each point in the (x, y) plane can be transformed into an (a, b, R) hyper-plane or accumulator plane.

To characterize the circle, a 3D accumulator plane with R, a and b is generated. For a specific (R, a, b) and for each θ value, the corresponding x and y value are computed using Equation 10.2. For every x and y value that is a foreground pixel i.e., the (x, y) value lies on the circle, a value of 1 is added to the specific (R, a, b) coordinate in the accumulator plane. This process is repeated for all values of (R, a, b).

The resultant accumulator hyper-plane will have high intensity at the points corresponding to a circle. Then the (R, a, b) corresponding to the local peak will provide the parameters of the circle in the original image.

The following is the Python function for the Hough circle transform:

```
cv2.HoughCircles(input, cv2.HOUGH_GRADIENT, dp,
  min_dist, param1, param2, minRadius, maxRadius);
```

```
Necessary argument:
input is a grayscale image as an ndarray.
```

```
cv2.HOUGH_GRADIENT is the method that is used by OpenCV.
```

```
  dp is the inverse ratio of resolution. If dp is an
integer n,
  then the accumulator width and height will be 1/n
  of the input image.
```

```
  min_dist is the minimum distance that the function
will maintain between the detected centers.
```

```
  param1 is the upper threshold for Canny edge detector
that is used by the Hough function internally.
```

```
  param2 is the threshold for center detection.
```

```
Optional arguments:
  min_radius is the minimum radius of the circle
  that needs to be detected while max_radius is
  the maximum radius to be detected.
```

Returns:

   output is an ndarray that contains information about
the (x, y) values of the center and radius of
each detected circle.

   The cv2 code for the Hough circle transform is given below.

```
import numpy as np
import scipy.ndimage
from PIL import Image
import cv2


# opening the image and converting it to grayscale
a = Image.open('../Figures/withcontrast1.png')
a = a.convert('L')
# Median filter is performed on the
# image to remove noise.
img = scipy.ndimage.filters.median_filter(a,size=5)
# Circles are determined using
# Hough circles transform.
circles = cv2.HoughCircles(img,
        cv2.HOUGH_GRADIENT,1,10,param1=100,
        param2=30,minRadius=10,maxRadius=30)
# circles image is rounded to unsigned integer 16.
circles = np.uint16(np.around(circles))
# For each detected circle.
for i in circles[0,:]:
# An outer circle is drawn for visualization.
    cv2.circle(img,(i[0],i[1]),i[2],(0,255,0),2)
# its center is marked
    cv2.circle(img,(i[0],i[1]),2,(0,0,255),3)
# Saving the image as houghcircles_output.png
cv2.imwrite('../Figures/houghcircles_output.png', img)
```

Figure 10.3(a) is a CT image with two bright white circular regions, which are contrast-filled blood vessels. The aim of this exercise is to characterize the vessel size using the Hough circle transform. The image is median filtered (Figure 10.3(b)) to remove noise. The size of the median filter kernel is 5-by-5. The search space is narrowed by specifying a minimum radius of 10 and a maximum radius of 30. The cv2.HoughCircles returns an ndarray with the inner array containing the center x, center y and radius respectively. The output of the Hough circle transform is shown in Figure 10.3(c). In the for-loop, the detected circles are marked using dark circles. The centers of these circles are also marked.

If the input image is of size N-by-N, the number of possible values of a and b are M and the number of possible values of R is K; the computational time is $O(KM^2N^2)$. Hence, the Hough circle transform is significantly computationally intensive compared to the Hough line transform. If the range of radii to be tested is smaller, then K is smaller and hence the computation can be made faster. If the approximate location of the circle is known, then the range of a and b is reduced and consequently decreases M and hence computation can be accomplished faster. Interested readers can refer to [IK88],[IK87],[LLM86],[Sha96] and [XO93] to learn more about Hough transforms.

10.4 Template Matching

The template matching technique is used to find objects in an image that match the given template. For example, template matching is used to identify a particular person in a crowd or a particular car in traffic. It works by comparing a sub-image of the person or object over a much larger image.

Template matching can be either intensity-based or feature-based. We will demonstrate intensity-based template matching. A mathe-

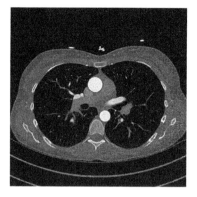
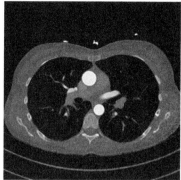

(a) Input Image.

(b) Image after applying median filter.

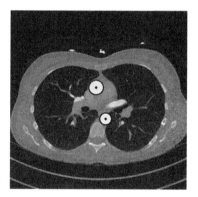

(c) Output with min radius = 10 and max radius 30.

FIGURE 10.3: An example of the Hough circle transform.

matical coefficient, called cross-correlation, is used for intensity-based template matching. Let $I(x,y)$ be the pixel intensity of image I at (x,y) then the cross-correlation, c between $I(x,y)$ and template $t(u,v)$ is given by

$$c(u,v) = \sum_{x,y} I(x,y)t(x-u,y-v) \qquad (10.3)$$

Cross-correlation is similar to the convolution operation. Since $c(u, v)$ is not independent of the changes in image intensities, we use the normalized cross-correlation coefficient proposed by J.P. Lewis [Lew95]. The normalized cross-correlation coefficient is given by the following equation:

$$r(u, v) = \frac{\sum_{x,y}(I(x, y) - \bar{I})(t(x - u, y - v) - \bar{t})}{\sqrt{\sum_{x,y}(I(x, y) - \bar{I})^2 \sum_{x,y}(t(x - u, y - v) - \bar{t})^2}} \qquad (10.4)$$

where $\bar{I}$ is the mean of the sub-image that is considered for template matching and $\bar{t}$ is the average of the template image. In the places where the template matches the image, the normalized cross-correlated coefficient is close to 1.

The following is the Python code for template matching.

```python
import cv2
import numpy
from PIL import Image
from skimage.morphology  import label
from skimage.measure import regionprops
from skimage.feature import match_template


# Opening the image and converting it to grayscale.
image = Image.open('../Figures/airline_seating.png')
image = image.convert('L')
# Converting the input image into an ndarray.
image = numpy.asarray(image)
# Reading the template image.
temp = Image.open('../Figures/template1.png')
temp = temp.convert('L')
# Converting the template into an ndarray.
```

```
temp = numpy.asarray(temp)
# Performing template matching.
result = match_template(image, temp)
thresh = 0.7
# Thresholding the result from template
# matching considering pixel values where the
# normalized cross-correlation is greater than 0.7.
res = result > thresh
# Labeling the thresholded image.
c = label(res, background=0)
# Performing regionprops to count the
# number of label.
reprop = regionprops(c)
print("The number of seats are:", len(reprop))
# Converting the binary image to an 8-bit for storing.
res = res*255
# Converting the ndarray to image.
cv2.imwrite("../Figures/templatematching_output.png", res)
```

The results of template matching are shown in Figure 10.4. Figure 10.4(a) is the input image containing the layout of airline seats and Figure 10.4(b) is the template image. The normalized cross-correlation coefficient, r, is computed for every pixel in the input image. Then the array comprising the normalized cross-correlated coefficients is thresholded. The threshold value of 0.7 is chosen. Then the regions in the thresholded array are labeled. Regionprops is performed on the labeled array to obtain the number of regions that match the template and have $r > 0.7$. The output image in Figure 10.4(c) is the thresholded image. In this particular example, the number of seats returned by the program is 263.

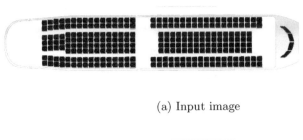

(a) Input image

(b) Template

(c) Cross-correlated image after segmentation

FIGURE 10.4: An example of template matching.

10.5   Corner Detector

A corner detector, as the name indicates, detects corners. It is typically a step for further image processing. For example, in medical imaging, the corner points could be used as an input for image registration, a process of transforming images from one coordinate system to another. Interested readers can refer to [Bir11] for more details on image registration.

We will discuss two corner detectors, namely, the FAST and Harris corner.

10.5.1 FAST Corner Detector

As the name indicates FAST corner detector ([RD06]) is computationally efficient. It works on the following principle.

1. Consider a pixel (p) in the image for corner detection and let its pixel value be v_p. We will use its neighboring 16 pixels in a circle for corner detection.

2. If a set of N pixels among the 16 pixels are either brighter or darker than the pixel p by a pre-determined threshold, the point is considered to be a corner.

3. Repeat this for all the pixels.

The computational complexity of this method is similar to convolution and hence relatively fast compared to other corner detectors.

In the code below, the image is opened and converted to a numpy array. The image is used to determine the response image using the corner_fast function. The corner_peaks function then finds the corner. Finally, a statistical test using corner_subpix is performed to ensure all the detected corners are corners. The detected corners are superimposed on the image for visualization.

```python
import numpy as np
from PIL import Image
from skimage.feature import corner_peaks
from skimage.feature import corner_subpix, corner_fast
from matplotlib import pyplot as plt

# Image is opened and is converted to grayscale.
img = Image.open('../Figures/corner_detector.png').
convert('L')
# img is converted to an ndarray.
img1 = np.asarray(img)
```

```
corner_response = corner_fast(img1)
cpv = corner_peaks(corner_response, min_distance=50)
corners_subpix_val = corner_subpix(img1, cpv,
    window_size=13)
fig, ax = plt.subplots()
ax.imshow(img1, interpolation='nearest', cmap=plt.cm.gray)
x = corners_subpix_val[:, 1]
y = corners_subpix_val[:, 0]
ax.plot(x, y, 'ob', markersize=10)
ax.axis('off')
plt.savefig('../Figures/corner_fast_detector_output.png',
    dpi=300)
plt.show()
```

Figure 10.5(a) is a segmented image of an electron microscopy sample. In the output image in Figure 10.5(b), the detected corners are highlighted using a star. As evidenced in the output, the FAST corner detector found spurious points while the Harris corner detector, which will be discussed next, produces fewer spurious points. In spite of this shortcoming, the FAST corner detector is useful in situations where speed is of essence. For example, for real-time corner detection, the FAST corner detector outperforms the Harris detector.

10.5.2   Harris Corner Detector

The Harris corner detector ([HS88]) works based on the following principle:

- The derivative of an image with no corners or edges along the x and y axes will be uniformly distributed.

- The derivative of an image with no corners but vertical edges will have a strong directional preference along the vertical direction.

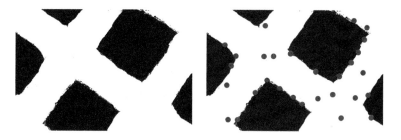

(a) Input image for the FAST corner detector.

(b) Output of the FAST corner detector.

FIGURE 10.5: An example of the FAST corner detector.

- The derivative of an image with no corners but horizontal edges will have a strong directional preference along the horizontal direction.

- The derivative of an image with corners will have a strong directional preference along both the vertical and horizontal directions.

We will begin the discusions by finding a difference image which is formed by finding the sum of the squared difference between a given pixel and its neighbors.

$$D(u, v) = \sum (I(x + u, y + v) - I(x, y))^2 \qquad (10.5)$$

For images with relatively similar pixel values nearby, the value of $D(u, v)$ will be zero. If there are significant changse in pixel values in the neighborhood of the pixel at position (x, y), then the value of $D(u, v)$ will be large. The aim in the Harris corner detector is to maximize the value of D.

We will simplify Equation 10.5 using Taylor series expansion and assuming the second and other higher partial derivatives can be ignored.

$$I(x + u, y + v) = I(x, y) + uI_x(x, y) + vI_y(x, y) \qquad (10.6)$$

where I_x and I_y are partial derivative of I along x and y axes.

By substituting 10.6 in 10.5, we obtain

$$D(u,v) = \sum u^2 I_x^2 + 2uv I_x I_y + v^2 I_y^2 \qquad (10.7)$$

which can be rewritten in matrix form as

$$D(u,v) = \sum \begin{pmatrix} u & v \end{pmatrix} \begin{pmatrix} I_x^2 & I_x I_y \\ I_x I_y & I_y^2 \end{pmatrix} \begin{pmatrix} u \\ v \end{pmatrix} \qquad (10.8)$$

which can be simplified to

$$D(u,v) = \begin{pmatrix} u & v \end{pmatrix} \left(\sum \begin{pmatrix} I_x^2 & I_x I_y \\ I_x I_y & I_y^2 \end{pmatrix} \right) \begin{pmatrix} u \\ v \end{pmatrix} \qquad (10.9)$$

which can be rewritten as

$$D(u,v) = \begin{pmatrix} u & v \end{pmatrix} M \begin{pmatrix} u \\ v \end{pmatrix} \qquad (10.10)$$

The Harris corner response (R) is then calculated as

$$R = det(M) - k(trace(M))^2 \qquad (10.11)$$

where *det* is the determinant of M and *trace* is the sum of all elements along the diagonal of M (i.e., trace of M) and k is a constant whose value range is from 0.04-0.06. R is a large value for corners while it is a small value for flat regions.

During the computation of gradients I_x and I_y, it is recommended to smooth the image using a Gaussian filter to reduce noise.

In the code below, the image is read and converted to a numpy array. The image is used to determine the response image using the corner_harris function. The corner_peaks image then finds the corner. Finally, a statistical test using corner_subpix is performed to ensure all the detected corners are corners.

```
import numpy as np
from PIL import Image
from matplotlib import pyplot as plt
from skimage.feature import corner_harris
from skimage.feature import corner_peaks, corner_subpix


# Opening image and converting it into grayscale.
img = Image.open('../Figures/corner_detector.png').
convert('L')
# img is converted to an ndarray.
img1 = np.asarray(img)


# Detecting corners using Harris.
corner_response = corner_harris(img1, k=0.2)
# Detecting peak values.
corners_peak_val = corner_peaks(corner_response, 50)


corners_subpix_val = corner_subpix(img1, corners_peak_val,
    13)
# Defining a subplot.
fig, ax = plt.subplots()
# Displaying the image.
ax.imshow(img1, interpolation='nearest', cmap=plt.cm.gray)
x = corners_subpix_val[:, 1]
y = corners_subpix_val[:, 0]
ax.plot(x, y, 'ob', markersize=10)
ax.axis('off')
# Saving the image.
plt.savefig('../Figures/corner_harris_detector_output.png',
    dpi=300)
plt.show()
```

Figure 10.6(a) is a segmented image of an electron microscopy sample. In the output image in Figure 10.6(b), the detected corners are highlighted using a star. As evidenced in the output, the Harris corner detector found fewer spurious points.

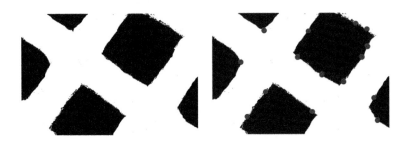

(a) Input image for the Harris corner detector.

(b) Output of the Harris corner detector.

FIGURE 10.6: An example of Harris corner detector.

10.6   Summary

- Labeling is used to identify different objects in an image.

- The regionprops function has several attributes and is used to study different properties of objects in a labeled image.

- The Hough line transform detects lines while the Hough circle transform detects circles. They also determine the corresponding parameters: slope and intercept for lines, and center and diameter for circles.

- Template matching is used to identify or count similar objects in an image.

- A corner detector is used to find corners in an image. It is typically used as a pre-processing step for further image processing. The FAST corner detector is computationally faster than the Harris corner detector but finds more spurious corners.

10.7 Exercises

1. The Hough transform is one method for finding the diameter of a circle. The process of finding the diameter is slow. Suggest a method for determining the **approximate diameter** of a circle, given only pixels corresponding to the two blood vessels in Figure 10.3(a).

2. Figure 4.9(a) in Chapter 4 consists of multiple characters. Write a Python program to break up this text and store the individual characters as separate images. Hint: Use the regionprops function.

3. Consider an image with 100 coins of various sizes spread on a uniform background. Assume that the coins do not touch each other, write a pseudo code to determine the number of coins for each size. Brave soul: Write a Python program to accomplish this. Hint: regionprops will be needed.

4. Consider an image with 100 coins of various sizes spread on a uniform background. Assume that the coins **do touch each other**, and write a pseudo code to plot a histogram of the area of the coin (along the x-axis) vs. the number of coins for a given area (along the y-axis). Write a Python program to accomplish this. If only a few coins overlap, determine the approximate number of coins.

Chapter 11

Neural Network

11.1 Introduction

Neural networks has taken the world by storm in the last decade. However, the work has been ongoing since the 1940s. Some of the initial work was in modeling the behavior of a biological neuron mathematically. Frank Rosenblat in 1958 built a machine that showed an ability to learn based on the mathematical notion of a neuron. The process of building neural networks were further refined over the next 4 decades. One of the most important papers that allowed training arbitrarily complex networks appeared in 1986 in a work by David E. Rumelhart, Geoffrey Hinton, and Ronald J. Williams [RHW86]. This paper re-introduced the back-propagation algorithm that is the workhorse of the neural network as used today. In the last 2 decades, due to the availability of cheaper storage and compute, large networks have been built that solve significant practical problems. This has made the neural network and its cousins such as the convolution neural network, recurrent neural network, etc., household names.

In this chapter, we will begin the discussion with the mathematics behind neural networks, which includes forward and back-propagation. We will then discuss the visualization of a neural network. Finally, we will discuss building a neural network using Keras, a Python module for machine learning and deep learning.

Interested readers are recommended to follow the discussions in the following sources: [Dom15], [MTH], [GBC16], [Gro17].

11.2   Introduction

A neural network is a non-linear function with many parameters. The simplest curve is a line with two parameters: slope and intercept. A neural network has many more parameters, typically in the order of 10,000 or more and sometimes millions. These parameters can be determined by the process of optimizing a loss function that defines the goodness of fit.

11.3   Mathematical Modeling

We will begin the discussion of the mathematics of a neural network by fitting lines and planes. We can then extend it to any arbitrary curve.

11.3.1   Forward Propagation

The equation of a line is defined as

$$y_1 = Wx + b \tag{11.1}$$

where x is the independent variable, y_1 is the dependent variable, W is the slope of the line and b is the intercept. In the world of machine learning, W is called the weight and b is the bias.

If the independent variable x is a scalar, then Equation 11.1 is a line, W is scalar and b is a scalar. However, if x is a vector, then Equation 11.1 is a plane, W is a matrix and b is a vector. If x is very large, Equation 11.1 is called a hyper-plane. Equation 11.1 is a linear equation and the best model that can ever be created using it would be a linear model as well.

In order to create non-linear models in a neural network, we add a non-linearity to this linear model. We will discuss one such non-linearity called sigmoid. In practice however, other non-linearities such as tanh, rectified linear unit (RELU), and leaky RELU are also used.

The equation of a sigmoid function is

$$y = \frac{1}{1 + e^{-x}} \tag{11.2}$$

When x is a large number, the value of y asymptotically reaches 1 (11.1), while for small values of x, the value of y asymptotically reaches 0. In the region along the x-axis between -1 and $+1$ approximately, the curve is linear and is non-linear everywhere else.

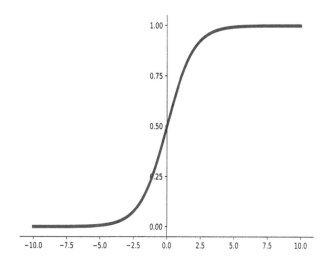

FIGURE 11.1: Sigmoid function.

If the y_1 from 11.1 is passed through a sigmoid function, we will obtain a new y_1,

$$y_1 = \frac{1}{1 + e^{-W_1 x - b_1}} \tag{11.3}$$

where W_1 is the weight of the first layer and b_1 is the bias of the first layer. The new y_1 in Equation 11.3 is a non-linear curve. The equation can be rewritten as,

$$y_1 = \sigma(W_1 x + b_1) \tag{11.4}$$

In a simple neural network here, we will add another layer (i.e., another set of W and b) to which we will pass the y_1 obtained from Equation 11.4.

$$y = W_2 y_1 + b_2 \tag{11.5}$$

where W_2 is the weight of the second layer and b_2 is the bias of the second layer.

If we substitute 11.4 in 11.5, we obtain,

$$y = W_2 \sigma(W_1 x + b_1) + b_2 \tag{11.6}$$

We can repeat this process by adding more layers and create a complex non-linear curve. However, for clarity sake, we will limit ourselves to 2 layers.

In Equation 11.6, there are 4 parameters namely, W_1, b_1, W_2, and b_2. If x is a vector, then W_1 and W_2 are matrices and b_1 and b_2 are vectors. The aim of a neural network is to determine the value inside these matrices and vectors.

11.3.2   Back-Propagation

The value of the 4 parameters can be determined using the process of back-propagation. In this process, we begin by assuming an initial value for the parameters. They can be assigned a value of 0 or some random value could be used.

We will then determine the initial value of y using Equation 11.6. We will denote this value as $\hat{y}$. The actual value y and the predicted value $\hat{y}$ will not be equal. Hence there will be an error between them. We will call this error "loss."

$$L = (\hat{y} - y)^2 \qquad (11.7)$$

Our aim is to minimize this loss by finding the correct value for the parameters of Equation 11.6.

Using the current value of the parameter, its new value iteratively can be calculated using,

$$W_{new} = W_{old} - \epsilon \frac{\partial L}{\partial W} \qquad (11.8)$$

where W is a parameter and L is the loss function. This equation is generally called an 'update equation'.

To simplify the calculation of partial derivatives such as $\frac{\partial L}{\partial W}$, we will derive them in parts and assemble them using chain rule.

We will begin by calculating $\frac{\partial L}{\partial \hat{y}}$ using Equation 11.7

$$\frac{\partial L}{\partial \hat{y}} = 2 * (\hat{y} - y) \qquad (11.9)$$

Then we will calculate $\frac{\partial L}{\partial W_2}$ using the chain rule,

$$\frac{\partial L}{\partial W_2} = \frac{\partial L}{\partial \hat{y}} * \frac{\partial \hat{y}}{\partial W_2} \qquad (11.10)$$

If we substitute $\hat{y}$ from Equation 11.5 and $\frac{\partial L}{\partial \hat{y}}$ from Equation 11.9, we obtain,

$$\frac{\partial L}{\partial W_2} = 2 * (\hat{y} - y) * y_1 \qquad (11.11)$$

The partial derivative can then be used to update the value of W_2 using the existing value of W_2 with the help of the update equation. A similar calculation (left as an exercise to the reader) can be shown for b_2 as well.

Next we will calculate the new value of W_1,

$$\frac{\partial L}{\partial W_1} = \frac{\partial L}{\partial \hat{y}} \frac{\partial \hat{y}}{\partial y_1} \frac{\partial y_1}{\partial W_1} \tag{11.12}$$

which can be computed using Equations 11.9, 11.5 and 11.4 respectively. Thus,

$$\frac{\partial L}{\partial W_1} = 2(\hat{y} - y)(W_2)(\sigma(W_1 x + b_1)(1 - \sigma(W_1 x + b_1))x) \tag{11.13}$$

which can be simplified to

$$\frac{\partial L}{\partial W_1} = 2x W_2(\hat{y} - y)\sigma(W_1 x + b_1)(1 - \sigma(W_1 x + b_1)) \tag{11.14}$$

The new value of W_1 can be calculated using the update Equation 11.8. A similar calculation (left as an exercise to the reader) can be shown for b_1 as well.

For every input data point or a batch of data points, we perform forward propagation, determine the loss, and then back-propagate to update the parameters (weights and biases) using the update equation. This process is repeated with all the available data.

In summary, the process of back-propagation finds the partial derivatives of the parameters of a neural network system and uses the update equation to find a better value for the parameters by minimizing the loss.

11.4 Graphical Representation

Typically, a neural network is represented as shown in Figure 11.2. The left layer is called the input layer, the middle is called the hidden layer, and the right is called the output layer.

A node (filled circle) in a given layer is connected to all the nodes in the next layer but is not connected to any nodes in that layer. In Figure 11.2, arrows are only drawn to originate from an input layer to the first node in the hidden layer. For clarity sake, the lines ending on other nodes are omitted. The values at each of the input nodes is multiplied with the weights in the line between the nodes. The weighted inputs are then added in the node in the hidden layer and passed through the sigmoid function or any other non-linearity. The output of the sigmoid function is then weighted in the next layer and the sum of all those weights will be the output of the output layer ($\hat{y}$).

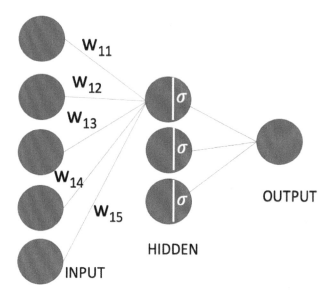

FIGURE 11.2: Graphical representation of a neural network.

If there are n nodes in the input layer and m nodes in the hidden layer, then the number of edges connecting from the input to hidden-layer will be n*m. This can be represented as a matrix of size [n, m]. Then the operation described in the previous paragraph will be a dot product between the input x and the matrix followed by application of the sigmoid function described in Equation 11.4. This matrix is the W_1 we have previously described.

If there are m nodes in the hidden layer and k nodes in the output layer, then the number of edges connecting from hidden to output will be m*k. This can be represented as a matrix of size [m, k] and is the matrix W_2 we have previously described.

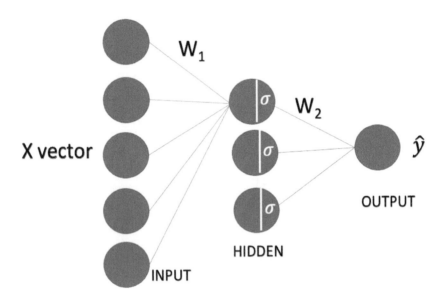

FIGURE 11.3: Graphical representation of a neural network as weight matrices.

During the forward propagation, a value of x is used as an input to begin the compute that is propagated from the input side to the output. The loss is calculated by comparing the predicted value and the actual. The gradients are then computed in reverse from the output layer toward the input and the parameters (weights and biases) are updated by back-propagation.

In this discussion, we assumed that y is a continuous function and its value is a real number. This class of problem is called a regression problem. An example of such a problem is the prediction of price of an item based on images.

11.5 Neural Network for Classification Problems

The other class of problem is the classification problem where the dependent variable y takes discrete values. An example of such a problem is identifying a specific type of lung cancer given an image. There are two major types: small cell lung cancer (SCLC) and non-small cell lung cancer (NSCLC).

In a classification problem, we aim to draw a boundary between two classes of points as shown in Figure 11.4. The two classes of points in the image are the circles and the plusses. A linear boundary (such as a line or plane) that has the lowest error cannot be drawn between these two sets of points. A neural network can be used to draw a non-linear boundary.

One of the common loss functions for the classification problem is the cross entropy loss. It is defined as

$$L = -\sum y \log \hat{y} \tag{11.15}$$

where y is the actual value and $\hat{y}$ is the predicted value.

Since the loss function is different compared to the regression problem, the derivatives such as $\frac{\partial L}{\partial W}$ would yield a different equation compared to the one derived for the regression problem. However the approach remains the same.

11.6 Neural Network Example Code

The current crop of popular deep learning packages such as Tensorflow [ABC+16], Keras [C+20], etc., require the programmer to define the forward propagation while the back-propagation is handled by the package.

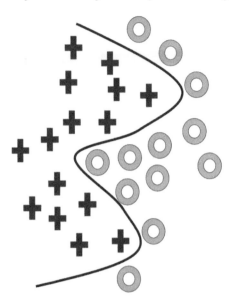

FIGURE 11.4: The neural network for the classification problem draws a non-linear boundary between two classes of points.

In the example below, we define a neural network to solve the problem of identifying handwritten digits from MNIST dataset [LCB10], a popular image dataset for benchmarking machine learning and deep learning applications. Figure 11.5 shows a few representative images from the MNIST dataset. Each image is 28 pixels by 28 pixels in size. The total number of pixels = 28*28 = 784. The images contain a single hand-drawn digit. As can be seen in the image, two numbers may not look the same in two different images. The task is to identify the digit in the image, given the image itself. The image is the input and the output is one of the 10 classes (number between 0 and 9).

We begin by importing all the necessary modules in Keras, specifically the Sequential model and Dense layer. The Sequential model allows defining a set of layers. In the mathematical discussion, we defined 2 layers. In Keras, these layers can be defined using the Dense layers class. A stack of these layers constitutes a Sequential layer.

We load the MNIST dataset using the convenient functionality (keras.datasets.mnist.load_data) available in Keras. This loads both the

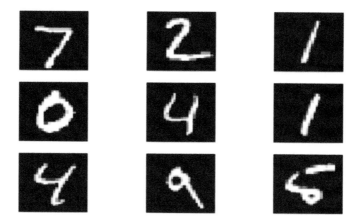

FIGURE 11.5: Some sample data from the MNIST dataset [LCB10].

training data as well as testing data. The number of images in the training dataset is 60,000 and the number of images in the testing dataset is 10,000. Each image is stored as a 784-pixel-long vector with 8-bit precision (i.e., pixel values are between 0 and 255). The corresponding y for each image is a single number corresponding to the digit in that image.

We then normalize the image by dividing each pixel value by 255 and subtracting 0.5. Hence the normalized image will have pixel values between −0.5 and +0.5.

The model is built by passing 3 Dense layers to the Sequential class. The first layer has 64 nodes, the second layer has 64 nodes. The first 2 layers use the Rectified Linear Unit (RELU) activation function for non-linearity. The last layer produces a vector of length 10. This vector is passed through a softmax function (Equation 11.16). The output of a softmax function is a probability distribution as each of the values corresponds to the probability of a given digit and also the sum of all the values in the vector equates to 1. Once we obtain this vector, determining the corresponding digit can be accomplished by finding the position in the vector with the highest probability value.

$$s_i = \frac{e^{x_i}}{\sum\limits_i e^{x_i}} \tag{11.16}$$

We will pass the model through an optimization process by calling the fit function. We run the model through 5 epochs, where each epoch is defined as visiting all images in the training dataset. Typically we feed a batch of images for training instead of one image at a time. In the example, we use a batch of 32, which implies in each training a random batch of 32 images and the corresponding labels are passed.

```python
import numpy as np
from keras.models import Sequential
from keras.layers import Dense
from keras.utils import to_categorical
from keras.datasets import mnist

# Fetch the train and test data.
(x_train, y_train), (x_test, y_test) = mnist.load_data()

# Normalize the image so that all pixel values
# are between -0.5 and +0.5.
x_train = (x_train / 255) - 0.5
x_test = (x_test / 255) - 0.5

# Reshape the train and test images to size 784 long vector.
x_train = x_train.reshape((-1, 784))
x_test = x_test.reshape((-1, 784))

# Define the neural network model with 2 hidden layer
# of size 64 nodes each.
model = Sequential([
    Dense(64, activation='relu', input_shape=(784,)),
    Dense(64, activation='relu'),
    Dense(10, activation='softmax'),
])
```

```
# Compile the model using Adam optimizer and use
# the cross entropy loss.
model.compile(optimizer='adam',
              loss='categorical_crossentropy',
              metrics=['accuracy'])


# Train the model.
model.fit(x_train, to_categorical(y_train), epochs=5,
          batch_size=32)
```

The output contains the result of training 5 epochs. As can be seen the value of cross entropy loss decreases as the training progresses. It started at 0.3501 and finally ended at 0.0975. Similarly, the accuracy increased as the training progressed from 0.8946 to 0.9697.

```
Epoch 1/5
60000/60000 [===] - 3s 58us/step - loss: 0.3501 - accuracy:
0.8946
Epoch 2/5
60000/60000 [===] - 3s 56us/step - loss: 0.1790 - accuracy:
0.9457
Epoch 3/5
60000/60000 [===] - 3s 55us/step - loss: 0.1357 - accuracy:
0.9576
Epoch 4/5
60000/60000 [===] - 3s 55us/step - loss: 0.1129 - accuracy:
0.9649
Epoch 5/5
60000/60000 [===] - 3s 57us/step - loss: 0.0975 - accuracy:
0.9697
```

Interested readers must consult the Keras documentation for more details.

11.7 Summary

- Neural networks are universal function approximators. In training a neural network, we fit a non-linear curve using available data.

- To obtain non-linearity in a neural network, we combine a linear function with non-linear functions such as sigmoid, RELU, etc.

- The parameters of the non-linear curve are learnt through the process of back-propagation.

- Neural networks can be used for both regression and classification problems.

11.8 Exercises

1. You are given a neuron that performs addition of y = x1*w1+x2*w2, where x1 and x2 are the inputs and w1 and w2 are weights. Write the back-propagation equation for it. Also write the update equation for w1 and w2.

2. In a neural network, we combine linear function $Wx + b$ with a non-linear function. We stack these layers together to produce an arbitrarily complex non-linear function. What would happen if we do not use a non-linear function but still stack layers? What kind of curve can we build?

3. Why is sigmoid no longer popular as an activation function? Conduct research on this topic.

Chapter 12

Convolutional Neural Network

12.1 Introduction

The convolution neural network (CNN) is a biologically inspired mathematical model of vision. The journey began successfully with the work by David Hubel and Torsten Wiesel who won the 1981 Nobel prize in Physiology or Medicine for this work. The work by Hubel and Weisel was best summarized by the Nobel committee's press release ([ppr20]) from 1981. The following paragraph is a reproduction from the press release:

"... the visual cortex's analysis of the coded message from the retina proceeds as if certain cells read the simple letters in the message and compile them into syllables that are subsequently read by other cells, which, in turn, compile the syllables into words, and these are finally read by other cells that compile words into sentences that are sent to the higher centers in the brain, where the visual impression originates and the memory of the image is stored."

As the quote indicates, Hubel and Wiesel found that the brain has a series of neurons. The neurons nearest to the retina detect simple shapes such as lines in different orientation. The neurons next to detect complex shapes like curves. The neurons downstream detect more complex shapes like nose, ear, etc.

The understanding of the brain's visual cortex paved the way for mathematical modeling of the visual pathway. The first successful work was done by Kunihiko Fukushima ([Fuk80]). He demonstrated a hierarchical model using convolution and downsampling. The convolution allowed viewing of only a part of the image or video while processing. The downsampling was performed by averaging. Many years later, a different method called "maxpooling" was introduced which is still in use today and will be discussed later in this chapter. The next major breakthrough was the work of Yann Lecun [LBD+89] who introduced a back-propagation approach to learn the parameters of a CNN.

With the availability of large quantities of data, cheaper storage, compute power and software, CNNs have become a go-to tool for solving image processing and computer vision problems in all areas of science and engineering.

12.2 Convolution

We discussed the process of applying convolution to an image in Chapter 4. In this section, we will discuss convolution from the perspective of a CNN. In the example in Chapter 4, the convolution was performed using a 5-by-5 filter where each element in the filter has a value of $\frac{1}{25}$. In a CNN, the values in the filter are determined by the learning process (i.e., the back-propagation process).

Also, unlike the examples from Chapter 4, in a CNN there are multiple filters used. These filters are arranged into layers. The first layer is designed to detect simple objects such as lines. Since there are many possible configurations (slopes) for lines, the first layer may have multiple filters to detect lines at all these orientations. The second layer is designed to detect curves. Since there are more configurations for a curve compared to a line, the number of filters in the second layer is typically more than the number of filters in the first layer. Modern CNNs[1] typically have more than 2 layers.

---

[1]CNNs are only 40 years old. We are distinguishing CNN architectures from the last 10 years from the ones by using the word modern.

12.3 Maxpooling

Maxpooling is a dimensionality reduction technique. It takes an input such as an image and reduces its size using the maximum among its neighbors. The effect of replacing a pixel value with its neighborhood maximum produces an abstracted representation of the image. We will demonstrate with an example.

Let's consider a small image (Figure 12.1(a)) of size 4x4 and also consider a maxpooling filter of size 2x2 placed on the top left corner of the image. In the maxpooling process, we will find the maximum value of the 2x2 region containing the values 10, 6, 8 and 2. We will create a new image (Figure 12.1(b)) where we will use the maximum value (10) from the previous calculation. We will then move this maxpooling filter by 2 steps (called a stride) over the image and find the next region on the image consisting of values 4, 6, 12 and 5. Its maximum value of 12 will be used for the next pixel in the output image. Once a row of maxpooling operation is completed, we move 2 rows below and continue this process.

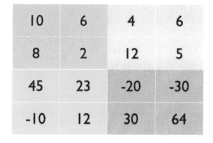

(a) Input image.

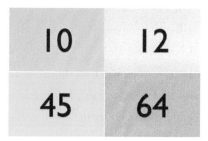

(b) Maxpooled image.

FIGURE 12.1: An example of applying maxpooling on a sub-image.

If an image is of size NxN and we move with a stride of 2x2, then the output image will be of size $\frac{N}{2}$x$\frac{N}{2}$. A higher stride can be used to reduce the image further.

The output from the convolution and maxpooling layer is finally passed to a classifier or a regressor, which is typically built using a neural network as discussed in the previous chapter. This is due to the fact that the convolution and maxpooling layer are data conditioners that prepare the data for a final classifier or a regressor.

12.4   LeNet Architecture

We will use the convolution layer and maxpooling to build LeNet [LBBH98], one of the first CNNs to revolutionize the field of computer vision. An input image (Figure 12.2) is passed to a series of 6 convolutions in the first layer. The output of the first convolution layer is subsampled using maxpooling, and is then passed to the second convolution layer, which contains 16 filters. The output of the second convolution layer is passed to the second maxpooling layer. The output of this layer is then flattened to a vector and passed to a neural network-based classifier or regressor.

Ideally, any classifier such as a Support Vector Machine SVM or Logistic regression can be used. However, a neural network, as discussed in the last chapter is preferred.

In the last chapter, we discussed that the parameters of the system can be learned using the process of back-propagation. The same applies to parameters (the values in the filter) of a CNN as well.

For every input image or a batch of images, we perform forward propagation through the convolution, maxpooling, and the neural network layers. We then determine the loss. Then we back-propagate through the neural network layers followed by back-propagation through the convolution layers and update the parameters (weights and biases) using the update equation. This process is repeated with all the available data.

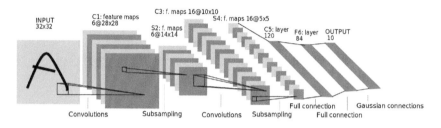

FIGURE 12.2: LeNet architecture diagram.

In the example below, we define a LeNet CNN to solve the problem of identifying handwritten digits from MNIST data set, which we also used in the last chapter.

We begin by importing all the necessary functionalities in Keras, specifically the Sequential model, Dense layer, Conv2D layer, and Max-Pooling2D layer. The Sequential model allows defining a set of layers. The list of layers in order is shown in Figure 12.3.

The first layer is the input layer that takes a 28x28x1 image. The second layer is the convolution layer with 32 filters. The third layer is the maxpooling layer that reduces the image size by $\frac{1}{2}$. The fourth and fifth layer are the second set of a convolution layer with 64 filters and a second maxpooling layer that reduces the image size by $\frac{1}{2}$. The second maxpooling layer output image is flattened and passed through two neural network layers to produce an output prediction.

In this example, we use images of size 28x28x1. After passing through the first convolution layer, we will obtain a 3D volume of size 28x28x32. The first maxpooling layer reduces the size by $\frac{1}{2}$ to 14x14x32. This volume is passed through the second convolution layer which produces a volume of size 14x14x64. The second maxpooling layer reduces the image to 7x7x64. The flattened vector will thus be of size 3136 which is the product of 7, 7 and 64.

In the code, we load the training and test dataset by using the 'mnist.load_data() method'. The x values (image pixels) are normalized

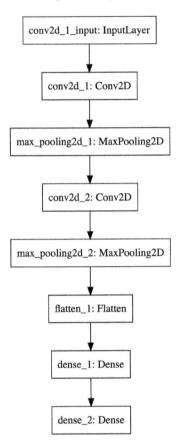

FIGURE 12.3: LeNet Keras model.

to be within the range $[-0.5, 0.5]$. They are then reshaped from its original shape of 50000x784 to 50000x28x28x1.

A sequential model is created and the various layers are added as described earlier. In all the layers, RELU non-linearity is added. The last layer is passed through softmax to obtain probability distribution which can be evaluated for cross-entropy loss. Finally, we evaluate the model for accuracy using the test data.

```
import numpy as np
import keras
from keras.datasets import mnist
```

```python
from keras.models import Sequential
from keras.layers import Dense, Flatten
from keras.layers import Conv2D, MaxPooling2D
from keras.utils import to_categorical


# Size of image is 28x28x1 channel.
input_shape = (28, 28, 1)
batch_size = 64
# number of possible outcomes [0-9]
nclasses = 10
epochs = 3


# Fetch the train and test data.
(x_train, y_train), (x_test, y_test) = mnist.load_data()


# Normalize the image so that all pixel values
# are between -0.5 and +0.5.
x_train = (x_train / 255) - 0.5
x_test = (x_test / 255) - 0.5


# Reshape the train and test images to size 28x28x1.
x_train = x_train.reshape((x_train.shape[0], *input_shape))
x_test = x_test.reshape((x_test.shape[0], *input_shape))


# Define the CNN model with 2 convolution layer and
# 2 max pooling layer followed by a neural network
# with 1 hidden layer of size 128 nodes.
model = Sequential()
model.add(Conv2D(32, kernel_size=(3, 3),
                 activation='relu',
                 input_shape=input_shape))
model.add(MaxPooling2D(pool_size=(2, 2)))
model.add(Conv2D(64, (3, 3), activation='relu'))
```

```
model.add(MaxPooling2D(pool_size=(2, 2)))
model.add(Flatten())
model.add(Dense(128, activation='relu'))
model.add(Dense(nclasses, activation='softmax'))


# Compile the model using Adam optimizer and use
# the cross entropy loss.
model.compile(optimizer='adam',
             loss='categorical_crossentropy',
             metrics=['accuracy'])


# Train the model.
model.fit(x_train, to_categorical(y_train), epochs=epochs,
         batch_size=batch_size)


# Evaluate the model.
score = model.evaluate(x_test, to_categorical(y_test),
    verbose=0)
print('Test loss:', score[0])
print('Test accuracy:', score[1])
```

In the case of CNN, in comparison to the neural network example from the previous chapter, you will notice that we reached a high accuracy value in fewer epochs although each epoch took a little longer.

```
Epoch 1/3
60000/60000 [====] - 33s 555us/step - loss: 0.1683 -
accuracy: 0.9504
Epoch 2/3
60000/60000 [====] - 27s 444us/step - loss: 0.0493 -
accuracy: 0.9847
Epoch 3/3
60000/60000 [====] - 48s 792us/step - loss: 0.0331 -
accuracy: 0.9898
```

```
Test loss: 0.03353308427521261
Test accuracy: 0.9894000291824341
```

12.5   Summary

- CNN was originally developed as a mathematical model of vision. Hence they are well suited for solving computer vision problems.

- CNNs are created by composition of convolution and maxpooling layer followed by a classifier or regressor, which is typically a neural network.

- The parameters of the convolution layer are learned using the back-propagation process.

12.6   Exercises

1. What is the effect of increasing the number of convolution layers in a neural network?

2. Modify the above code and run it with the FashionMNIST data set available at `https://github.com/zalandoresearch/fashion-mnist`. This data set also has 10 categories such a trousers, shoes, etc.

Part III

Image Acquisition

Chapter 13

X-Ray and Computed Tomography

13.1 Introduction

So far we have covered the basics of Python, its scientific modules, and image processing techniques. In this chapter, we begin our journey of learning image acquisition. We will begin the discussion with x-ray generation and detection. We will discuss the various modes in which x-ray interacts with matter. These methods of interaction and detection have resulted in many modes of x-ray imaging such as angiography, fluoroscopy, etc. We complete the discussion with the basics of CT, reconstruction, and artifact removal.

13.2 History

X-rays were discovered by Wilhelm Conrad Röntgen, a German physicist, during his experiment with cathode ray tubes. He called these mysterious rays "x-rays," the symbol "x" being used in mathematics to denote unknown variables. He found that unlike visible light, these rays passed through most materials and left a characteristic shadow on a photographic plate. His work was published as "On New Kind of Rays" [R95] and was subsequently awarded the first Nobel Prize in Physics in 1901.

Subsequent study of x-rays revealed their true physical nature. They are a form of electromagnetic radiation similar to light, radio waves, etc. They have a wavelength of 10 to 0.01 nanometers. Although they are well known and studied and no longer mysterious, they continue to be referred to as x-rays. Even though the majority of x-rays are man-made using x-ray tubes, they are also found in nature. The branch of x-ray astronomy studies celestial objects by measuring the x-rays emitted.

Since Röntgen's days, the x-ray has found very widespread use across various fields including radiology, geology, crystallography, astronomy, etc. In the field of radiology, x-rays are used in fluoroscopy, angiography, computed tomography (CT), etc. Today, many non-invasive surgeries are performed under x-ray guidance, providing a new "eye" to the surgeons.

13.3   X-Ray Generation

An x-ray imaging system consists of a generator producing a constant and reliable output of x-rays, an object (typically a patient) through which the x-ray traverses, and an x-ray detector to measure the intensity of the rays after passing through the object. We begin with a discussion of the x-ray generation process using an x-ray tube.

13.3.1   X-Ray Tube Construction

An x-ray tube consists of four major parts. They are an anode, a cathode, a tungsten target, and an evacuated tube to hold the three parts together, as shown in Figure 13.1.

The cathode (negative terminal) produces electrons (negatively charged) that are accelerated toward the anode (positive terminal). The filament in the cathode is heated by passing current, which generates electrons by the process of thermionic emission, defined as emission

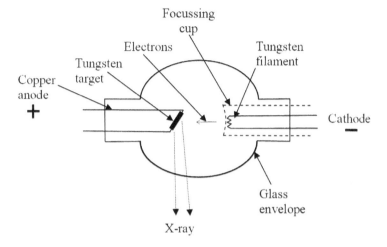

FIGURE 13.1: Components of an x-ray tube.

of electrons by absorption of thermal energy. The number of electrons produced is proportional to the current through the filament. This current is generally referred to as "tube current" and is generally measured in "mA" or "milli-amperes."

Since the interior of an x-ray tube can be hot, a metal with a high melting point such as tungsten is chosen for the filament. Tungsten is also a malleable material, ideal for making fine filaments. The electron produced is focused by the focusing cup, which is maintained at the same negative potential as the cathode. The glass enclosure in which the x-ray is generated is evacuated so that the electrons do not interact with other molecules and can also be controlled independently and precisely. The focusing cup is maintained at a very high potential in order to accelerate the electrons produced by the filament.

The anode is bombarded by the fast-moving electrons. The anode is generally made from copper so that the heat produced by the bombardment of the electrons can be properly dissipated. A tungsten target is fixed to the anode. The fast-moving electrons either knock out the electrons from the inner shells of the tungsten target or are slowed due to the tungsten nucleus. The former results in the characteristic x-ray spectrum while the latter results in the general spectrum or

Bremsstrahlung spectrum. The two spectrums together determine the energy distribution in an x-ray and will be discussed in detail in the next section.

The cathode is stationary but the anode can be stationary or rotating. The rotating anode allows even distribution of heat and consequently longer life of the x-ray tube.

There are three parameters that control the quality and quantity of an x-ray. These parameters together are sometimes referred to as an x-ray technique.

They are:

1. Tube voltage measured in kVp.

2. Tube current measured in mA.

3. X-ray exposure time in ms.

In addition, a filter (such as a sheet of aluminum) is placed in the path of the beam, so that lower-energy x-rays are absorbed. This will be discussed in the next section.

The tube voltage is the electric potential between the cathode and the anode. Higher voltage results in increased velocity of the electrons between the cathode and the anode. This increased velocity will produce high-energy x-rays while a lower voltage results in lower-energy x-rays and consequently a noisier image. The tube current determines the number of electrons being emitted. This in turn determines the quantity of x-rays. The exposure time determines the time for which the object or patient is exposed to x-rays. This is generally the time the x-ray tube is operating.

13.3.2   X-Ray Generation Process

The x-ray generated by the tube does not contain photons of a single energy. It instead consists of a large range of energy. The relative number of photons at each energy level is measured to generate a histogram. This histogram is called the spectral distribution or spectrum for short. There are two types of x-ray spectrums [CDM84b]. They are the general radiation or Bremsstrahlung "Braking" spectrum, which is a continuous radiation, and the characteristic spectrum, a discrete entity as shown in Figure 13.2.

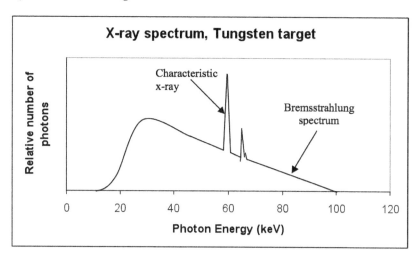

FIGURE 13.2: X-ray spectrum illustrating characteristic and Bremsstrahlung.

When the fast-moving electrons produced by the cathode move very close to the nucleus of the tungsten atom (Figure 13.3), the electrons decelerate and the loss of energy is emitted as radiation. Most of the radiation is at a higher wavelength (or lower energy) and hence is dissipated as heat. The electrons are not decelerated completely by one tungsten nucleus, and hence at every stage of deceleration, radiation of lower wavelength or higher energy is emitted. Since the electrons are decelerated or "braked" in the process, this spectrum is referred to as Bremsstrahlung or braking spectrum. This spectrum gives the x-ray spectrum its wide range of photon energy levels.

From the energy equation, we know that

$$E = \frac{hc}{\lambda} \tag{13.1}$$

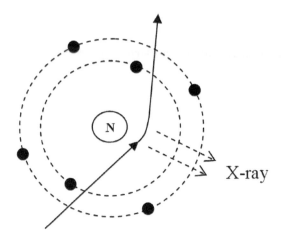

FIGURE 13.3: Production of Bremsstrahlung or braking spectrum.

where $h = 4.135 * 10^{-18} eVs$ is Planck's constant, $c = 3 * 10^8 m/s$ is the speed of light, and λ is the wavelength of the x-ray measured in Angstroms (Å$= 10^{-10}m$). The product of h and c is $12.4 * 10^{-10} keV m$. When E is measured in keV, the equation simplifies to

$$E = \frac{12.4}{\lambda} \qquad (13.2)$$

The inverse relationship between E and λ implies that a shorter wavelength produces a higher-energy x-ray and vice versa. For an x-ray tube powered at 112 kVp, the maximum energy that can be produced is 112 keV and hence the corresponding wavelength is 0.11 Å. This is the shortest wavelength and also the highest energy that can be achieved during the production of Bremsstrahlung spectrum. This is the right-most point in the graph in figure 13.2. However, most of the x-ray will be produced at much higher wavelength and consequently lower energy.

The second type of radiation spectrum (Figure 13.4) results from a tungsten electron in its orbit interacting with the emitted electron. This is referred to as characteristic radiation, as the peaks in the histogram of the spectrum are characteristic of the target material.

The fast-moving electrons eject the electron from the k-shell (inner shell) of the tungsten atom. Since this shell is unstable due to the ejection of the electron, the vacancy is filled by an electron from the outer shell. This is accompanied by release of x-ray energy. The energy and wavelength of the electron are dependent on the binding energy of the electron whose position is filled. Depending on the shell, these characteristic radiations are referred as K, L, M and N characteristic radiation and are shown in Figure 13.2.

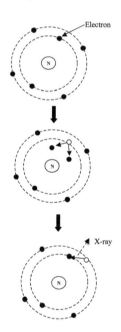

FIGURE 13.4: Production of characteristic radiation.

X-rays do not just interact with the tungsten atom, they can interact with any atom in their path. Thus, a molecule of oxygen in the path will be ionized by an x-ray knocking out its electron. This could change the x-ray spectrum and hence the x-ray generator tube is maintained at vacuum.

13.4   Material Properties

13.4.1   Attenuation

Once the x-ray is generated, it is allowed to pass through a patient or an object. The material in the object reduces the intensity of the x-ray either by absorption or deflection of photons in the beam. This process is referred to as attenuation. If there are multiple materials, each of the materials can absorb or deflect the x-ray and consequently reduce its intensity.

The attenuation is quantified by using a linear attenuation coefficient (μ), defined as the attenuation per centimeter of the object. The attenuation is directly proportional to the distance traveled and the incident intensity. The intensity of the x-ray beam after attenuation is given by the Lambert-Beer law (Figure 13.5) expressed as

$$I = I_0 e^{-\mu \delta x} \tag{13.3}$$

where I_0 is the initial x-ray intensity, I is the exiting x-ray intensity, μ is the linear attenuation coefficient of the material, and δx is the thickness of the material. The law also assumes that the input x-ray intensity is mono-energetic or monochromatic.

Monochromatic radiation is characterized by photons of single intensity, but in reality all radiations are polychromatic and have photons of varying intensity with spectra similar to Figure 13.2. Polychromatic radiation is characterized by photons of varying energy (quality and quantity), with the peak energy being determined by the peak kilovoltage (kVp).

When polychromatic radiation passes through matter, the longer wavelengths and lower energy are preferentially absorbed. This increases the mean energy of the beam. This process of increased mean energy of the beam is referred to as "beam hardening."

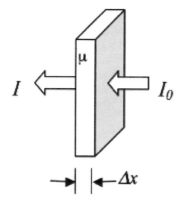

FIGURE 13.5: Lambert-Beer law for monochromatic radiation and for a single material.

In addition to the attenuation coefficient, the characteristics of a material under x-ray can also be defined using the half-value layer. This is defined as the thickness of material needed to reduce the intensity of the x-ray beam by half. So from Equation 13.3 for a thickness $\delta x = HVL$ (half value layer),

$$I = \frac{I_0}{2} \tag{13.4}$$

Hence,

$$I_0 e^{-\mu HVL} = \frac{I_0}{2} \tag{13.5}$$

$$\mu HVL = 0.693 \tag{13.6}$$

$$HVL = \frac{0.693}{\mu} \tag{13.7}$$

For a material with linear attenuation coefficient of 0.1/cm, the HVL is 6.93 cm. This implies that when a monochromatic beam of x-ray passes through the material, its intensity drops by half after passing through 6.93 cm of that material.

TABLE 13.1: Relationship between kVp and HVL.

kVp	HVL(mm of Al)
50	1.9
75	2.8
100	3.7
125	4.6
150	5.4

The HVL depends not only on the material being studied but also on the tube voltage. High tube voltage produces a smaller number of low-energy photons, i.e., the spectrum in Figure 13.2 will be shifted to the right. The mean energy will be higher and the beam will be harder. This hardened beam can penetrate material without a significant loss of energy. Thus, the HVL will be high for high x-ray tube voltage. This trend can be seen in the HVL of aluminum at different tube voltages given in Table 13.1.

13.4.2 Lambert-Beer Law for Multiple Materials

For an object with n materials (Figure 13.6), the Lambert-Beer law is applied in cascade,

$$I = I_0 e^{-\mu_1 \Delta x} e^{-\mu_2 \delta x} ... e^{-\mu_n \delta x} = I_0 e^{-\sum_{i=1}^{n} \mu_i \Delta x} \qquad (13.8)$$

When the logarithm of the intensities is taken, for a continuous domain we obtain

$$p = -\ln\left(\frac{I}{I_0}\right) = \sum_{i=1}^{n} \mu_i \Delta x = \int \mu(x)dx \qquad (13.9)$$

Using this equation, we see that the value p, the projection image expressed in energy intensity, corresponding to the digital value at a specific location in that image, is simply the sum of the product of attenuation coefficients and thicknesses of the individual components.

This is the basis of image formation in x-ray. The inverse of this summation process is the CT reconstruction that will be discussed shortly.

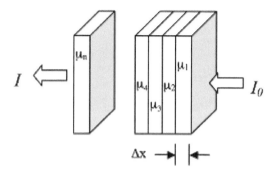

FIGURE 13.6: Lambert-Beer law for multiple materials.

13.4.3 Factors Determining Attenuation

The energy of the beam is one of the factors that determines the amount of attenuation. As we have seen earlier, lower-energy beams get preferentially absorbed compared to higher-energy beams.

The density of a substance through which x-rays pass makes a significant contribution to the attenuation. A higher-density substance like bone attenuates x-rays more than a lower-density substance like tissue. Also different types of tissue have different densities and hence different attenuation, resulting in different contrast on the x-ray image. The physical characteristic that determines the attenuation is the number of electrons per gram in the material. A material with a higher number of electrons per gram has a higher probability of interacting with the x-rays. The number of electrons per gram is given by

$$N = \frac{N_0 Z}{A} \tag{13.10}$$

where N is the number of electrons per gram, $N_0 = 6 * 10^{23}$ is Avogadro's number, Z is the atomic number, and A is the atomic weight of the substance. Since Avogadro's number is a constant, the number of electrons per gram is dependent only on Z and A.

13.5   X-Ray Detection

So far, we have discussed x-ray generation using an x-ray tube, the shape of the x-ray spectrum, and also studied the change in x-ray intensity as it traverses a material due to attenuation. These attenuated x-rays have to be converted to a human-viewable form. This conversion process can be achieved by exposing them on a photographic plate to obtain an x-ray image or viewing them using a TV screen or converting to a digital image, all using the process of x-ray detection. There are three different types of x-ray radiation detectors in practice, namely ionization, fluorescence, and absorption.

1. Ionization detection

   In the ionization detector, the x-rays ionize the gas molecules in the detector and by measuring the ionization, the intensity of the x-ray is measured. An example of such a detector is the Geiger Muller counter [Mac83] shown in Figure 13.7. These detectors are used to measure the intensity of radiation and are not used for creating x-ray images.

2. Scintillation detection

   There are different types of scintillation detectors. The most popular are the Image Intensifier (II) and Flat Panel Detector (FPD). In an II, [Mac83], [CDM84b], [FH00], the x-rays are converted to electrons that are accelerated to increase their energy. The electrons are then converted back to light and are viewed on a TV or a computer screen. In the case of an FPD, the x-rays are converted to visible light and then to electrons using a photo diode.

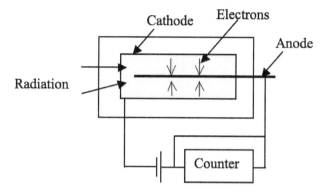

FIGURE 13.7: Ionization detector.

The electrons are recorded using a camera. In both the II and FPD, the process of converting x-ray to electron and accelerating it is used for improving the image gain. Modern technology has allowed the creation of a large FPD with very high quality and hence the FPD is rapidly replacing the II. Also, FPD occupies significantly less space than the II. We will discuss both in detail.

13.5.1   Image Intensifier

The II (Figure 13.8) consists of an input phosphor and photocathode, an electrostatic focusing lens, an accelerating anode and an output fluorescent screen. The x-ray beam passes through the patient and enters the II through the input phosphor. The phosphor generates light photons after absorbing the x-ray photons. The light photons are absorbed by the photocathode and electrons are emitted. The electrons are then accelerated by a potential difference toward the anode. The anode focuses the electron onto an output fluorescence screen that emits the light that will be displayed using a TV screen, recorded on an x-ray film, or recorded by a camera onto a computer.

The input phosphor is made of cesium iodide (CsI) and is vapor deposited to form a needlelike structure that prevents diffusion of light and hence improves resolution. It also has greater packing density and hence higher conversion efficiency, even with smaller thickness (needed for good spatial resolution). A photocathode emits electrons when light photons are incident on it. The anode accelerates the electrons. The higher the acceleration the better is the conversion of electrons to light photons at the output phosphor. The input phosphor is curved, so that electrons travel the same length toward the output phosphor. The output fluorescent screen is silver-activated zinc-cadmium sulfide. The output can be viewed using a series of lenses on a TV or it can be recorded on a film.

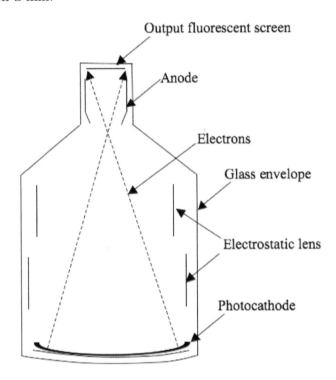

FIGURE 13.8: Components of an image intensifier.

13.5.2 Multiple-Field II

The field size is changed by changing the position of the focal point, the point of intersection for the left and right electron beams. This is achieved by increasing the potential in the electrostatic lens. Lower potential results in the focus being close to the anode and hence the full view of the anatomy is exposed to the output phosphor. At higher potential, the focus moves away from the anode and hence only a portion of the input phosphor is exposed to the output phosphor. In both cases, the size of the input and output phosphors remains the same but in the smaller mode, a portion of the image from the input phosphor is removed from the view due to a farther focal point.

In a commercial x-ray unit, these sizes are specified in inches. A 12-inch mode will cover a larger anatomy while a 6-inch mode will cover a smaller anatomy. Exposure factors are automatically increased for smaller II modes to compensate for the decreased brightness from minification.

Since the electrons travel large distances during their journey from photocathode to anode, they are affected by the earth's magnetic field. The earth's magnetic field changes even for small motions of the II and hence the electron path gets distorted. The distorted electron path produces a distorted image on the output fluorescent screen. The distortion is not uniform but increases near the edge of the II. Hence the distortion is more significant for a large II mode than for a smaller II mode. The distortions can be removed by careful design and material selection or more preferably using image processing algorithms.

13.5.3 Flat Panel Detector (FPD)

The FPD (Figure 13.9) consists of a scintillation detector, a photo diode, an amorphous silicon, and a camera. The x-ray beam passes through the patient and enters the FPD through the scintillation detector. The detector generates light photons after absorbing the x-ray photons. The light photons are absorbed by the photo diode and electrons are emitted. The electrons are then absorbed by the amorphous silicon layer, which produces an image that is recorded using a charge-couple device (CCD) camera.

Similar to the II, the scintillation detector is made of cesium iodide (CsI) or gadolinium oxysulfide and is vapor deposited to form a needle-like structure, which acts like fiber optic cable and prevents diffusion of light and improves resolution. The CsI is generally coupled with amorphous silicon, as CsI is an excellent absorber of x-ray and emits light photons at a wavelength best suited for amorphous silicon to convert to electrons.

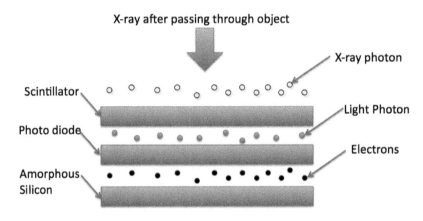

FIGURE 13.9: Flat panel detector schematic.

The II needs extra length to allow accelerating of the electron, while the FPD does not. Hence the FPD occupies significantly less space compared to the II. The difference becomes significant as the size of the detector increases. IIs are affected by the earth's magnetic field while such problems do not exist for the FPD. Hence the FPD can be mounted on an x-ray machine and be allowed to rotate around the patient without distorting the image. Although the II suffers from some disadvantages, it is simpler in its construction and electronics.

The II or FPD can be bundled with an x-ray tube, a patient table, and a structure to hold all these parts together, to create an imaging system. Such a system could also be designed to revolve around the patient table axis and provide images in multiple directions to aid diagnosis or medical intervention. Examples of such systems, fluoroscopy and angiography, are discussed below.

13.6   X-Ray Imaging Modes

13.6.1   Fluoroscopy

The first-generation fluoroscope [Mac83],[CDM84b] consisted of a fluoroscopic screen made of copper-activated cadmium sulfide that emitted light in the yellow-green spectrum of visible light. The image was so faint that the viewing was carried out in a dark room, with the doctors adapting their eyes to the dark prior to examination. Since the intensity of fluorescence was less, rod vision in the eye was used and hence the ability to differentiate shades of gray was also poor. These problems were alleviated with the invention of the II discussed earlier. The II allowed intensification of the light emitted by the input phosphor so that it could safely and effectively be used to produce a system (Figure 13.10) that could generate and detect x-rays and also produce images that can be studied using TVs and computers.

13.6.2   Angiography

A digital angiographic system [Mac83], [CDM84b] consists of an x-ray tube, a detector such as a II or FPD and a computer to control the system and record or process the images. The system is similar to fluoroscopy except that it is primarily used to visualize blood vessels opacified using a contrast. The x-ray tube must have a larger focal spot and also provide a constant output over time. The detector must also provide a constant acceleration voltage to prevent variation in gain during acquisition. A computer controls the whole imaging chain.

The computing system also performs digital subtraction in the case of digital subtraction angiography (DSA) [CDM84b] on the obtained images. In the DSA process, the computer controls the x-ray technique so that uniform exposure is obtained across all images. The computer obtains the first set of images without the injection of contrast and stores them as a mask image. Subsequent images obtained under the

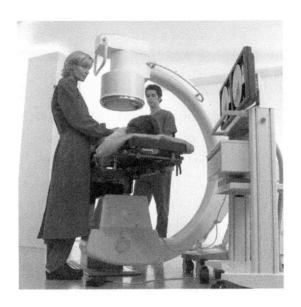

(a) Fluoroscopy machine.

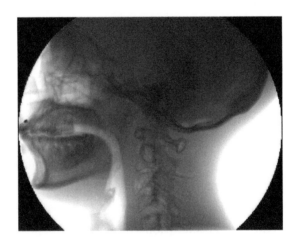

(b) Image of a head phantom acquired using a
II system.

FIGURE 13.10: Fluoroscopy machine. Original image reprinted with permission from Siemens AG.

injection of contrast are stored and subtracted from the mask image to obtain the image with the blood vessel alone.

13.7 Computed Tomography (CT)

The fluoroscopy and angiography discussed so far produce a projection image, which is a shadow of part of the body under x-ray. These systems provide a planar view from one direction and may also contain other organs or structures that impede the ability to make a clear diagnosis. CT on the other hand, provides a slice through the patient and hence offers an unimpeded view of the organ of interest. In CT, a series of x-ray images are acquired all around the object or patient. A computer then processes these images to produce a map of the original object using a process called reconstruction. Sir Godfrey N. Hounsfield and Dr. Allan McCormack developed CT independently and later shared the Nobel Prize for Physiology in 1979. The utility of this technique became so apparent that an industry quickly developed around it, and it continues to be an important diagnostic tool for physicians and surgeons. For more details refer to [Bus00],[Hen83],[Kal00].

13.7.1 Reconstruction

The basic principle of reconstruction is that the internal structure of an object can be computed from multiple projections of that object. In the case of CT reconstruction, the internal structure being reconstructed is the spatial distribution of the linear attenuation coefficients (μ) of the imaged object. Mathematically, Equation 13.9 can be inverted by the reconstruction process to obtain the distribution of the attenuation coefficients.

In clinical CT, the raw projection data is often a series of 1D vectors of x-ray projection obtained at various angles for which the 2D reconstruction yields a 2D attenuation coefficient matrix. In the case of 3D CT, a series of 2D images obtained at various angles are used to obtain a 3D distribution of the attenuation coefficient. For the sake of simplicity, the reconstructions discussed in this chapter will focus on 2D reconstructions and hence the projection images are 1D vector unless otherwise specified.

13.7.2 Parallel-Beam CT

The original method used for acquiring CT data used parallel-beam geometry such as is shown in Figure 13.11. As shown in the figure, the paths of the individual rays of x-ray from the source to the detector are parallel to each other. An x-ray source is collimated to yield a single x-ray beam, and the source and detector are translated along the axis perpendicular to the beam to obtain the projection data (a single 1D vector for a 2D CT slice). After the acquisition of one projection, the source-detector assembly is rotated and subsequent projections are obtained. This process is repeated until a 180-degree projection is obtained. The reconstruction is obtained using the central slice theorem or the Fourier slice theorem [KS88]. This method forms the basis for many CT reconstruction techniques.

13.7.3 Central Slice Theorem

Consider the object shown in Figure 13.12 to be reconstructed. The original coordinate system is x-y and when the detector and x-ray source are rotated by an angle θ, then their coordinate system is defined by $x' - y'$. In this figure, R is the distance between the iso-center (i.e., center of rotation) and any ray passing through the object. After logarithmic conversion, the x-ray projection at an angle (θ) is given by

$$g_\theta(R) = \int \int f(x, y)\delta(x \cos \theta + y \sin \theta - R)dx \, dy \qquad (13.11)$$

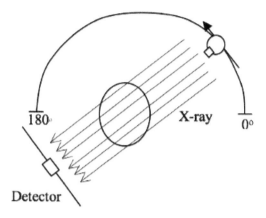

FIGURE 13.11: Parallel-beam geometry.

where δ is the Dirac-Delta function [Bra99].

The Fourier transform of the distribution is given by

$$F(u, v) = \int \int f(x, y)e^{-i2\pi(ux+vy)}dx\ dy \qquad (13.12)$$

where u and v are frequency components in perpendicular directions. Expressing u and v in polar coordinates, we obtain $u = \nu\cos\theta$ and $v = \nu\sin\theta$, where ν is the radius and θ is the angular position in the Fourier space.

Substituting for u and v and simplifying yields,

$$F(\nu, \theta) = \int \int f(x, y)e^{-i2\nu\pi(x\cos\theta + y\sin\theta)}dx\ dy \qquad (13.13)$$

The equation can be rewritten as

$$F(\nu, \theta) = \int \int \int f(x, y)e^{-i2\pi\nu R}\delta(x\cos\theta + y\sin\theta - R)dR\ dx\ dy \qquad (13.14)$$

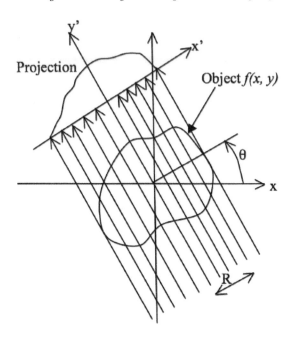

FIGURE 13.12: Central slice theorem.

Rearranging the integrals yields,

$$F(\nu, \theta) = \int \left(\int \int f(x, y) \delta(x \cos \theta + y \sin \theta - R) \right) e^{-i2\pi \nu R} dR$$

$$(13.15)$$

From Equation 13.11, we can simplify the above equation as

$$F(\nu, \theta) = \int g_e(R) e^{i2\pi \nu R} dR = FT(g_e(R)) \qquad (13.16)$$

where FT( ) refers to the Fourier transform of the enclosed function. Equation 13.16 shows that the radial slice along an angle θ in the 2D Fourier transform of the object is the 1D Fourier transform of the projection data acquired at that angle θ. Thus, by acquiring projections at various angles, the data along the radial lines in the 2D Fourier transform can be obtained. Note that the data in the Fourier space is obtained using polar sampling. Thus, either a polar inverse Fourier transform must be performed or the obtained data must be interpolated onto a rectilinear Cartesian grid so that Fast Fourier Transform (FFT) techniques can be used.

However, another approach can also be taken. Again, $f(x, y)$ is related to the inverse Fourier transform, i.e.,

$$f(x, y) = \int \int F(\nu, \theta)e^{i2\pi(ux+vy)} du\ dv \tag{13.17}$$

By using a polar coordinate transformation, u, v can be written as $u = \cos\theta$ and $v = \sin\theta$. To effect a coordinate transformation, the Jacobian is used and is given by

$$J = \begin{vmatrix} \frac{\partial u}{\partial \nu} & \frac{\partial u}{\partial \theta} \\ \frac{\partial v}{\partial \nu} & \frac{\partial v}{\partial \theta} \end{vmatrix} = \begin{vmatrix} \cos\theta & -\nu\sin\theta \\ \sin\theta & \nu\cos\theta \end{vmatrix} = \nu \tag{13.18}$$

Hence,

$$du\ dv = |\nu|d\nu\ d\theta \tag{13.19}$$

Thus,

$$f(x, y) = \int \int F(\nu, \theta)e^{i2\pi(x\cos\theta+y\sin\theta)}|\nu|d\nu\ d\theta \tag{13.20}$$

Using Equation 13.16, we can obtain

$$f(x, y) = \int \int FT(g_\theta(R))e^{i2\pi(x\cos\theta+y\sin\theta)}|\nu|d\nu\ d\theta \tag{13.21}$$

$$f(x, y) = \int \int FT(g_\theta(R))e^{i2\pi\nu R}\delta(x\cos\theta + y\sin\theta - R)|\nu|d\nu\ d\theta\ dR \tag{13.22}$$

$$f(x, y) = \int \int \left(FT(g_\theta(R))|\nu|e^{i2\pi\nu R}d\nu\right)\delta(x\cos\theta + y\sin\theta - R)d\theta\ dR \tag{13.23}$$

The term in the braces is the filtered projection, which can be obtained by multiplying the Fourier transform of the projection data by $|\nu|$ in the Fourier space or equivalently by performing a convolution of the real space projections and the inverse Fourier transform of the function $|\nu|$. Because the function looks like a ramp, the filter generated is commonly called the "ramp filter."

Thus,

$$f(x,y) = \int \int FT(R,\theta) \bullet \delta(x\cos\theta + y\sin\theta - R)d\theta\ dR \qquad (13.24)$$

where $FT(R,\theta)$ is the filtered projection data at location R acquired at angle θ is given by

$$f(x,y) = \int \int FT(g_\theta(R))|\nu|e^{i2\pi\nu R}dR \qquad (13.25)$$

Once the convolution or filtering is performed, the resulting data is reconstructed using Equation 13.25. This process is referred to as the filtered back projection (FBP) technique and is the most commonly used technique in practice.

13.7.4   Fan-Beam CT

The fan-beam CT scanners (Figure 13.13) have a bank of detectors, with all detectors being illuminated by x-rays simultaneously from every projection angle. Since the detector acquires images in one x-ray exposure, it eliminates the translation at each angle. Since translation is eliminated, the system is mechanically stable and faster. However, x-rays scattered (we will discuss scatter correction later) by the object reduce the contrast in the reconstructed images compared to parallel-beam reconstruction. But these machines are still popular due to faster acquisition time, which allows reconstruction of a moving object, like slices of the heart, in one breath-hold. The images acquired using fan-beam scanners can be reconstructed using a rebinning method that converts fan-beam data into parallel-beam data and then uses the central slice theorem for reconstruction. Currently, this approach is not used and is replaced by a direct fan-beam reconstruction method based on filtered back-projection.

A fan-beam detector with one row of detecting elements produces one CT slice. The current generations of fan-beam CT machines have multiple detector rows and can acquire 8, 16, 32 slices, etc., in one rotation of the object and are referred to as multi-slice CT machines. The benefit is faster acquisition time compared to single slice and also covering a larger area in one exposure. With the advent of multi-slice CT machines, a whole-body scan of the patient can also be obtained.

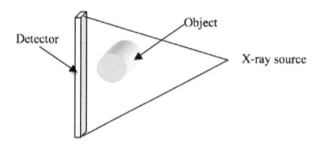

FIGURE 13.13: Fan-beam geometry.

Figure 13.14 is the axial slice of the region around the human kidney. It is one of the many slices of the whole body scan shown in the montage in Figure 13.15. These slices were converted into a 3D object (Figure 13.16) using Mimics^{TM} [Mat20a].

13.7.5 Cone-Beam CT

Cone-beam acquisition or CBCT (Figure 13.17) consists of 2D detectors instead of 1D detectors used in the parallel and fan-beam acquisitions. As with fan-beam, the source and detector rotate relative to the object, and the projection images are acquired. The 2D projection images are then reconstructed to obtain 3D volume. Since a 2D region is imaged, cone-beam-based volume acquisition makes use

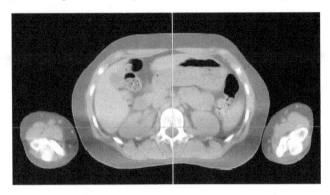

FIGURE 13.14: Axial CT slice.

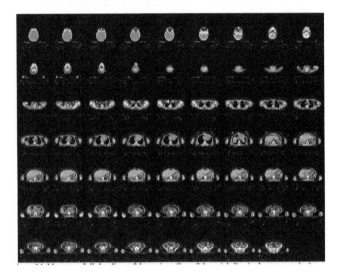

FIGURE 13.15: Montage of all the CT slices of the human kidney region.

of x-rays that otherwise would have been blocked. The advantages are potentially faster acquisition time, better pixel resolution, and isotropic (same voxel size in x, y and z directions) voxel resolution. The most commonly used algorithm for cone-beam reconstruction is the Feldkamp algorithm [FDK84], which assumes a circular trajectory for the source and flat detector and is based on filtered back-projection.

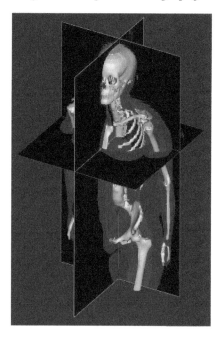

FIGURE 13.16: 3D object created using the axial slices shown in the montage. The 3D object in green is superimposed on the slice information for clarity.

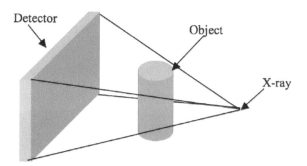

FIGURE 13.17: Cone beam geometry.

13.7.6 Micro-CT

Micro-tomography (commonly known as industrial CT scanning), like tomography, uses x-rays to create cross-sections of a 3D object that later can be used to re-create a virtual model without destroying the original model. The term micro is used to indicate that the pixel sizes of the cross-sections are in the micrometer range. These pixel sizes have also resulted in the terminology micro-computed tomography, micro-CT, micro-computer tomography, high-resolution x-ray tomography, and similar terminologies. All of these names generally represent the same class of instruments.

This also means that the machine is much smaller in design compared to the human version and is used to image smaller objects. In general, there are two types of scanner setups. In the first setup, the x-ray source and detector are typically stationary during the scan while the animal or specimen rotates. In the second setup, much more like a clinical CT scanner, the animal or specimen is stationary while the x-ray tube and detector rotate.

The first x-ray micro-CT system was conceived and built by Jim Elliott in the early 1980s [ED82]. The first published x-ray micro-CT images were reconstructed slices of a small tropical snail, with pixel size about 50 micrometers, which appeared in the same paper.

Micro-CT is generally used for studying small objects such as polymers, plastics, micro devices, electronics, paper, and fossils. It is also used in the imaging of small animals such as mice, or insects, etc.

13.8   Hounsfield Unit (HU)

The HU is the system of units used in CT that represents the linear attenuation coefficient of an object. It provides a standard way of comparing images acquired using different CT machines. The conversion of reconstructed pixel values to HUs is a linear transformation given by

$$HU = \left(\frac{\mu - \mu_w}{\mu_w} \right) * 1000 \qquad (13.26)$$

where μ is the linear attenuation coefficient of the object and μ_w is the linear attenuation coefficient of water. Thus, water has an HU of 0 and air has an HU of -1000 since μ of air is 0.

The following are the steps to obtain the HU equivalent of a reconstructed image:

- A water phantom consisting of a cylinder filled with water is reconstructed using the same x-ray technique as the reconstructed patient slices.

- The attenuation coefficient of water and air (present outside the cylinder) is measured from the reconstructed slice.

- A linear fit is established with the HU of water (0) and air (-1000) being the ordinate and the corresponding linear attenuation coefficients measured from the reconstructed image being the abscissa.

- Any patient data reconstructed is then mapped to HUs using the determined linear fit.

Since the CT data is calibrated to HUs, the data in the images acquires meaning not only qualitatively but also quantitatively. Thus, an HU number of 1000 for a given pixel or voxel represents quantitatively a bone in an object.

Unlike MRI, microscopy, ultrasound, etc., due to use of HUs for calibration, CT measurement is a map of a physical property of the material. This is handy while performing image segmentation, as the same threshold or segmentation technique can be used for measurements from various patients at various intervals and conditions. It is also useful in performing quantitative CT, a process of measuring the property of the object using CT.

13.9   Artifacts

In all the prior discussions, it was assumed that the x-ray beam is mono-energetic. It was also assumed that the geometry of the imaging system is well characterized, i.e., there is no change in the orbit that the imaging system follows with reference to the object. However, in current clinical CT technology, the x-ray beam is not mono-energetic and the geometry is not well characterized. This results in errors in the reconstructed image that are commonly referred as artifacts, defined as any discrepancy between the reconstructed value in the image and the true attenuation coefficients of the object [Hsi03]. Since the definition is broad and can encompass many things, discussions of artifacts are generally limited to clinically significant errors. CT is more prone to artifacts than conventional radiography, as multiple projection images are used. Hence errors in different projection images cumulate to produce artifacts in the reconstructed image. These artifacts could annoy radiologists or in some severe cases hide important details that could lead to misdiagnosis.

Artifacts can be eliminated to some extent during acquisition. They can also be removed by pre-processing projection images or post-processing the reconstructed images. There are no generalized techniques for removing artifacts and hence new techniques are devised depending on the application, anatomy, etc. Artifacts cannot be completely eliminated but can be reduced by using correct techniques, proper patient positioning, and improved design of CT scanners, or by software provided with the CT scanners.

There are many sources of error in the imaging chain that can result in artifacts. They can generally be classified as artifacts due to the imaging system or artifacts due to the patient. In the following discussion, the geometric alignment, offset and gain correction are caused by the imaging system while the scatter and beam-hardening artifacts are caused by the nature of object or patient being imaged.

13.9.1 Geometric Misalignment Artifacts

The geometry of a CBCT system is specified using six parameters, namely the three rotation angles (angles corresponding to u, v and w axes in Figure 13.18) and three translations along the principal axis (u, v, w in Figure 13.18). An error in these parameters can result in a ring artifact [CMSJ05],[FH00], double wall artifact, etc., which are visual and hence cannot be misdiagnosed as a pathology. However, very small errors in these parameters can result in blurring of edges and hence misdiagnosis of the size of the pathology, or shading artifacts that could shift the HU number. Hence these parameters must be determined accurately and corrected prior to reconstruction.

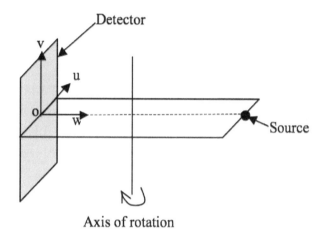

FIGURE 13.18: Parameters defining a cone-beam system.

13.9.2 Scatter

In our previous discussion, we learned that an incident x-ray photon ejects an electron from the orbit of the atom, and consequently a low-energy x-ray photon is scattered from the atom. The scattered photon travels at an angle from its incident direction (Figure 13.19). These scattered radiations are detected but arrive at the detector like the primary radiation. They reduce the contrast of the image and produce blurring. The effect of scatter in the final image is different for conventional radiography and CT. In the case of radiography, the images have poor contrast but in the case of CT, the logarithmic transformation results in a non-linear effect.

Scatter also depends on the type of image acquisition technique. For example, fan-beam CT has less scatter compared to a cone-beam CT due to the smaller height of the beam.

One of the methods to reduce scatter is the air gap technique. In this technique, a large air-gap is maintained between the patient and the detector. Since the scattered radiation at a large angle from the incident direction cannot reach the detector, it will not be used in the formation of the image. It is not always possible to provide an air gap between the patient and the detector, so grids or post-collimators [CDM84b],[Hsi03] made of lead strips are used to reduce scatter. The grids contain space which corresponds to the photo-detector being detected. The scattered radiation arriving at a large angle will be absorbed by lead and only primary radiations arriving at a small angle from the incident direction is detected. The third approach is software correction [LK87],[OFKR99]. Since scatter is a low-frequency structure causing blurring, it can be approximated by a number estimated using the beam-stop technique [Hsi03]. This, however, does not remove the noise associated with the scatter.

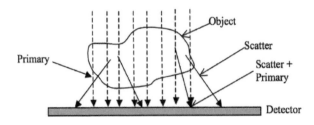

FIGURE 13.19: Scatter radiation.

13.9.3 Offset and Gain Correction

Ideally, the response of a detector must remain constant for a constant x-ray input at any time. But due to temperature fluctuations during acquisition, non-idealities in the production of detectors and variations in the electronic readouts, a non-linear response may be obtained in the detectors. These non-linear responses result in the output of that detector cell being inconsistent with reference to all the neighboring detector pixels. During reconstruction, the non-linear responses produce ring artifacts [Hsi03] with their center being located at the iso-center. These circles may not be confused with human anatomy, as there are no parts which form a perfect circle, but they degrade the quality of the image and hide details and hence need to be corrected. Moreover the detector produces some electronic readout, even when the x-ray source is turned off. This readout is referred to as "dark current" and needs to be removed prior to reconstruction.

Mathematically the flat field and zero offset corrected image (IC) is given by

$$IC(x,y) = \frac{IA - ID}{IF - ID}(x,y) * Average(IF - ID) \qquad (13.27)$$

where IA is the acquired image, ID is the dark current image, IF is flat field image, which is acquired at the same technique as the acquired image with no object in the beam. The ratio of the differences is multiplied by the average value of $(IF - ID)$ for gain normalization. This process is repeated for every pixel. The dark field images are to be acquired before each run, as they are sensitive to temperature variations. Other software-based correction techniques based on image processing are also used to remove the ring artifacts. They can be classified as pre-processing and post-processing techniques. The pre-processing techniques are based on the fact that the rings in the reconstructed images appear as vertical lines in the projection space. Since no feature in an object except those at the iso-center can appear as vertical lines, the pixels corresponding to vertical lines can be replaced using estimated pixel values. Even though the process is simple, the noise and complexity of human anatomy present a big challenge in the detection of vertical lines. Another correction scheme is the post-processing technique [Hsi03]. The rings in the reconstructed images are identified and removed. Since ring detection is primarily an arc detection technique, it could result in over-correcting the reconstructed image for features that look like arcs. So in a supervised ring removal technique, inconsistencies across all views are considered. To determine the position of pixels corresponding to a given ring radius, a mapping that depends on the location of source, object and image is used.

13.9.4 Beam Hardening

The spectrum (Figure 13.2) does not have a unique energy but has a wide range of energies. When such an energy spectrum is incident on a material, the lower energy gets attenuated faster as it is preferentially absorbed than the higher-energy. Hence a polychromatic beam becomes harder or richer in higher-energy photons as it passes through the material. Since the reconstruction process assumes an "ideal" monochromatic beam, the images acquired using a polychromatic beam produce cupping artifacts [BK04]. The cupping artifact is characterized by a radial increase in intensity from the center of the reconstructed image to its periphery. Unlike ring artifacts, this artifact presents difficulty, as it can mimic some pathology and hence can lead to misdiagnosis. The cupping artifacts also shift the intensity values and hence present difficulty in quantification of the reconstructed image data. They can be reduced by hardening the beam prior to reaching the patient, using a filter made of aluminum, copper, etc. Algorithmic approaches [BK04] for reducing these artifacts have also been proposed.

13.9.5   Metal Artifacts

Metal artifacts are caused by the presence of materials that have a high attenuation coefficient when compared to pathology in the human body. These include surgical clips, biopsy needles, tooth fillings, implants, etc. Due to their high attenuation coefficient, metal artifacts produce beam-hardening artifacts (Figure 13.20) and are characterized by streaks emanating from the metal structures. Hence techniques used for removing beam hardening can be used to reduce these artifacts.

In Figure 13.20, the top image is a slice taken at a location without any metal in the beam. The bottom image contains an applicator. The beam hardening causes a streaking artifact that not only renders the metal poorly reconstructed but also adds streaks to nearby pixels and hence makes diagnosis difficult.

Algorithmic approaches [Hsi03], [JS78], [WSOV96] to reducing these artifacts have been proposed. A set of initial reconstructions is performed without any metal artifact correction. From the reconstructed image, the location of metal objects is then determined. These objects are then removed from the projection image to obtain a synthesized projection. The synthesized projection is then reconstructed to obtain a reconstructed image without metal artifacts.

13.10   Summary

- A typical x-ray and CT system consists of an x-ray tube, detector and a patient table.

- X-ray is generated by bombarding high-speed electrons on a tungsten target. A spectrum of x-ray is generated. There are two parts to the spectrum: Bremsstrahlung or braking spectrum and the characteristic spectrum.

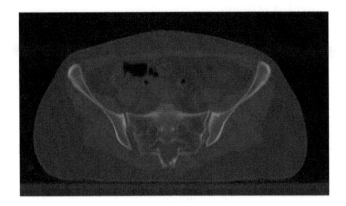

(a) Slice with no metal in the beam

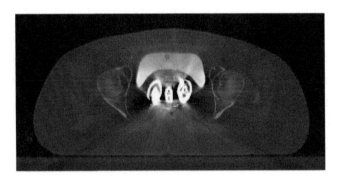

(b) Beam hardening artifact with strong streaks emanating from a metal applicator

FIGURE 13.20: Effect of metal artifact.

- The x-ray passes through a material and is attenuated. This is governed by the Lambert-Beer law.

- The x-ray after passing through a material is detected using either an ionizing detector or a scintillation detector such as the II or FPD.

- X-ray systems can be either fluoroscopic or angiographic.

- A CT system consists of an x-ray tube and detector, that are rotated around the patient to acquire multiple images. These images are reconstructed to obtain the slice through a patient.

- The central slice theorem is an analytical technique for reconstructing images. Based on this theorem, it can be proven that the reconstruction process consists of filtering and then back-projection.

- The Hounsfield unit is the unit of measure in CT. The unit is a map of the attenuation coefficient of the material.

- CT systems suffer from various artifacts such as misalignment artifact, scatter artifact, beam hardening artifact, and metal artifact.

13.11 Exercises

1. Describe briefly the various parameters that control the quality of x-ray or CT images.

2. An x-ray tube has an acceleration potential of 50kVp. What is the wavelength of the x-ray?

3. Describe the difference in the detection mechanism between II and FPD. Specifically describe the advantages and disadvantages.

4. Allan M. Cormack and Godfrey N. Hounsfield won the 1979 Nobel Prize for creation of CT. Read their Nobel acceptance speech and understand the improvement in contrast and spatial resolution of the images described compared to current clinical images.

5. What is the HU value of a material whose linear attenuation coefficient is half of the linear attenuation coefficient of water?

6. A metal artifact causes significant distortion of an image both in structure and HU value. Using the list of papers in the references, summarize the various methods.

Chapter 14

Magnetic Resonance Imaging

14.1 Introduction

Magnetic Resonance Imaging (MRI) is built on the same physical principle as Nuclear Magnetic Resonance (NMR), which was first described by Dr. Isidor Rabi in 1938 and for which he was awarded the Nobel Prize in Physics in 1944. In 1952, Felix Bloch and Edward Purcell won the Nobel Prize in Physics for demonstrating the use of the NMR technique in various materials.

It took a few more decades to apply the NMR principle to imaging the human body. Paul Lauterbur developed the first MRI machine that generated 2D images. Peter Mansfield expanded on Paul Lauterbur's work and developed mathematical techniques that are still part of MRI image creation. For their work, Peter Mansfield and Paul Lauterbur were awarded the Nobel Prize in Physics in 2003.

MRI has developed over the years as one of the most commonly used diagnostic tools by physicians all over the world. It is also popular due to the fact that it does not use ionizing radiation. It is superior to CT for imaging tissue, due to its better tissue contrast.

Unlike the physics and workings of CT, MRI physics is more involved and hence this chapter is arranged differently than the CT chapter. In the x-ray and CT chapter, we began with the construction and generation of x-ray, then discussed the material properties that govern x-ray imaging, and finally discussed x-ray detection and image formation. However, in this chapter we begin the discussion with the various laws that govern NMR and MRI. This includes Faraday's law of electromagnetic induction, Larmor frequency, and the Bloch equation. This is followed by material properties such as T_1 and T_2 relaxation times and the gyromagnetic ratio and proton density that govern MRI imaging. This is followed by sections on NMR detection and finally MRI imaging. With all the physics understood, we discuss the construction of an MRI machine. We conclude with the various modes and potential artifacts in MRI imaging. Interested readers can find more details in [Bus88], [CDM84a], [DKJ06], [Hor95], [Mac83], [MW98], [McR03], [Spl10], [Wes09].

14.2   Laws Governing NMR and MRI

14.2.1   Faraday's Law

Faraday's law is the basic principle behind electric motors and generators. It is also part of today's electric and electric-hybrid cars. It was discovered by Michael Faraday in 1831 and was correctly theorized by James Clerk Maxwell. It states that current is induced in a coil at a rate at which the magnetic flux changes. In Figure 14.1, when the magnet is moved in and out of the coil in the direction shown, a current is induced in the coil in the direction shown. This is useful for electrical power generation, where the flux of the magnetic field is achieved by rotating a powerful magnet inside the coil. The power for the motion is obtained using mechanical means such as potential energy of water (hydroelectric), chemical energy of diesel (diesel engine power plants), etc.

The converse is also true. When a current is passed through a closed circuit coil, it will cause the magnet to move. By constricting the motion of the magnet to rotation, an electric motor can be created. By suitably wiring the coils, an electric generator can thus become an electric motor.

In the former, the magnet is rotated to induce current in the coil while in the latter, a current passed through the coil rotates the magnet.

MRI and NMR use electric coils for excitation and detection. During the excitation phase, the current in the coil will induce a magnetic field that causes the atoms to align in the direction of the magnetic field. During the detection phase, the change in the magnetic field is detected by measuring the induced current.

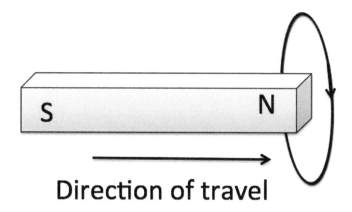

FIGURE 14.1: Illustration of Faraday's law.

14.2.2 Larmor Frequency

An atom (although a quantum-level object) can be described as a spinning top. Such a top will be precessing about its axis at an angle as shown in Figure 14.2. The frequency of the precession is an important factor and is described by the Larmor equation (Equation 14.1).

$$f = \gamma B \qquad (14.1)$$

where γ is the gyromagnetic ratio, f is the Larmor frequency, and B is the strength of the external magnetic field.

Magnetic Field

FIGURE 14.2: Precessing of the nucleus in a magnetic field.

14.2.3 Bloch Equation

An atom in a magnetic field is aligned in the direction of the field. An RF pulse (to be introduced later) can be applied to change the orientation of the atom. If the magnetic field is pointing in the z-direction, the atom will be aligned in the z-direction. If a pulse of sufficient strength is applied, the atom can be oriented in the x-direction or y-direction or sometimes even in the z-direction, the direction opposite to its original.

If the RF pulse is removed, the atom returns to its original z-direction. During the process of moving from the xy-direction to the z-direction, the atom traces a spiral motion, described by the Bloch equations (Equation 14.2).

$$
\begin{aligned}
M_x &= e^{-\frac{t}{T_2}}\cos\omega t \\
M_y &= e^{-\frac{t}{T_2}}\sin\omega t \\
M_z &= M_0(1 - e^{-\frac{t}{T_1}})
\end{aligned}
\tag{14.2}
$$

The equations can be easily visualized by plotting them in 3D (Figure 14.3). At time $t = 0$, the value of M_z is zero. This is due to the fact that the atoms are oriented in the xy-plane and hence their net magnetization is also in the xy-plane and not in the z-direction. When the RF pulse is removed, the atoms begin to orient in the z-direction (their original direction before RF pulse). This change along the xy-plane is an exponential decay in amplitude change and sinusoidal in directional change. Thus, the net magnetization reduces over time exponentially while sinusoidally changing direction in the xy-plane. At $t = infinity$, the M_x and M_y reach 0 while M_z reaches the original value of M_0.

Since a patient cannot be imaged for an infinite amount of time, in real applications, the atom will never recover its original magnetization completely.

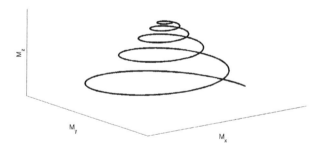

FIGURE 14.3: Bloch equation as a 3D plot.

14.3   Material Properties

14.3.1   Gyromagnetic Ratio

The gyromagnetic ratio of a particle is the ratio of its magnetic dipole moment to its angular momentum. It is a constant for a given nuclei. The values of the gyromagnetic ratio for various nuclei are given in Table 14.1. When an object containing multiple materials (and hence

TABLE 14.1: An abbreviated list of the nuclei of interest to NMR and MRI imaging and their gyromagnetic ratios.

Nuclei	γ (MHz/T)
H^1	42.58
P^{31}	17.25
Na^{23}	11.27
C^{13}	10.71

different nuclei) is placed in a magnetic field of certain strength, the precessional frequency is directly proportional to the gyromagnetic ratios based on the Larmor equation. Hence, if we measure the precessional frequency, we can distinguish the various materials. For example, the gyromagnetic ratio is 42.58 MHz/T for a hydrogen nucleus while it is 10.71 MHz/T for a carbon nucleus. For a typical clinical MRI machine, the magnetic field strength (B) is 1.5T. Hence the precessional frequency of a hydrogen atom is 63.87 MHz and that of the carbon is 16.07 MHz.

14.3.2 Proton Density

The second material property that is imaged is the proton density or spin density. It is the number of "mobile" hydrogen nuclei in a given volume of the sample. The higher the proton density, the larger the response of the sample in NMR or MRI imaging.

The response to NMR and MRI is not only dependent on the density of the hydrogen nucleus but also its configuration. A hydrogen nucleus connected to oxygen responds differently compared to the one connected to a carbon atom. Also, a tightly bound hydrogen atom does not produce any noticeable signal. The signal is generally produced by an unbound or free hydrogen nucleus. Thus, the hydrogen atom in tissue that is loosely bound produces a stronger signal. Bone on the other hand has hydrogen atoms that are strongly bound and hence produce a weaker signal.

TABLE 14.2: List of biological materials and their proton or spin density.

Biological material	Proton or spin density
Fat	98
Grey matter	94
White matter	100
Bone	1-10
Air	< 1

Table 14.2 lists the proton density of common materials. It can be seen from the table that the proton density for bone is low compared to white matter. Thus, the bone responds poorly to the MRI signal. One exception in the table is the response of fat to the MRI signal. Although fat consists of a large number of protons, it responds poorly to the MRI signal. This is due to the long chain of molecules found in fat that immobilize hydrogen atoms.

14.3.3 T_1 and T_2 Relaxation Times

There are two relaxation times that characterize the various regions in an object and can help distinguish them in an MRI image. They characterize the response of an atom in the Bloch equation.

Consider the image shown in Figure 14.4. A strong magnetic field B_0 is applied in the direction of the z-axis. This causes a net magnetization of M_0 in the z-axis to increase from zero. The increase is initially rapid but then slows down. It is given by Equation 14.3 and graphically represented by Figure 14.5.

$$M_z = M_0(1 - e^{-\frac{t}{T_1}}) \qquad (14.3)$$

The time for the net magnetization to reach a value within e (i.e., $M_0 - M_z = \frac{M_0}{e}$) is called the T_1 relaxation time. Since T_1 deals with both magnetization and demagnetization along the longitudinal direction (z-axis), T_1 is also referred to as **longitudinal relaxation time**.

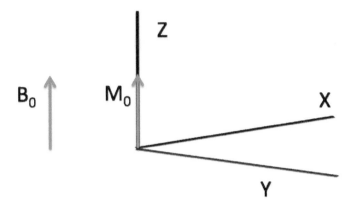

FIGURE 14.4: T_1 magnetization.

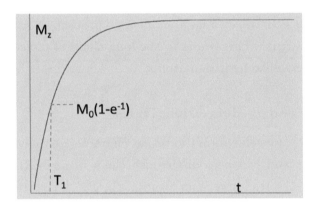

FIGURE 14.5: Plot of T_1 magnetization.

During an MRI image acquisition, in addition to the external magnetic field, an RF pulse is applied. This RF pulse disturbs the equilibrium and reduces M_z. The protons are not in isolation from other atoms but instead are bound tightly by a lattice. When the RF pulse is removed, the protons return to equilibrium, which causes a decrease in M_{xy} or transverse magnetization. This is accomplished by transferring energy to other atoms and molecules in the lattice. The time constant for the magnetization decay in the xy-axis is called T_2 or the

TABLE 14.3: List of biological materials and their T_1 and T_2 values for field strength of 1.0 T.

Biological material	T_1 (ms)	T_2 (ms)
Cerebrospinal fluid	2160	160
Grey matter	810	100
White matter	680	90
Fat	240	80

spin-lattice relaxation time. It is governed by Equation 14.4 and is graphically represented by Figure 14.6.

$$M_{xy} = M_{xy0}e^{-\frac{t}{T_2}} \tag{14.4}$$

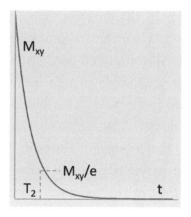

FIGURE 14.6: Plot of T_2 de-magnetization.

T_1 and T_2 are independent of each other but T_2 is generally smaller than or equal to T_1. This is evident from Table 14.3, which lists T_1 and T_2 values for some common biological materials. The value of T_1 and T_2 are dependent on the strength of the external magnetic field (1.0T in this case).

14.4   NMR Signal Detection

As discussed earlier, the presence of a strong magnetic field aligns the proton in the object in the direction of the magnetic field. The most interesting phenomenon happens when an RF pulse is applied to the object in the presence of the main magnetic field.

Due to the strong magnetic field (B_0), the protons align themselves with it. They also precess at the Larmor frequency. This is the equilibrium state of the proton under the magnetic field. When an RF pulse is applied using the transmitting coil to the cartoon head (Figure 14.7), the proton orientation changes and in some cases it flips in the negative direction while precessing at the Larmor frequency. Due to the flip, the net magnetization is in the direction opposite to the direction of the main magnetic field. When the RF pulse is removed, the protons flip back to the positive direction and hence reach their equilibrium state. During this process, an electric current is induced in the receiving coil due to changing magnetic field. This based on Faraday's law which was discussed previously. The signal obtained in the receiving coil is shown in Figure 14.8. The signal reduces in its intensity over time due to free induction decay (FID) and the time for the protons to reach their equilibrium state, or the "relaxed" state, is called the "relaxation time."

The signal is a plot over time. This signal contains details of the frequencies of various protons in the object. The frequency distribution can be obtained by using Fourier transform.

14.5   MRI Signal Detection or MRI Imaging

In this section, we will learn methods for obtaining images using MRI. The actual imaging process begins with selection of a section of

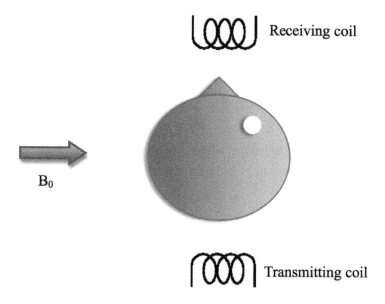

FIGURE 14.7: Net magnetization and effect of RF pulse.

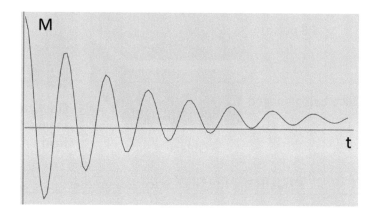

FIGURE 14.8: Free induction decay.

the object being imaged and placing that section under a magnetic field in a process called slice selection. An MRI signal can only be achieved by changing the orientation of the proton under the magnetic field. This is achieved by applying an RF pulse in the other two orthogonal directions during phase and frequency encoding. All these activities need to be timed so that an MRI image can be obtained. This timing process is called the pulse sequence. We will discuss each of these in detail in the subsequent sections.

14.5.1   Slice Selection

Slice selection is achieved by applying the magnetic field shown in figure 14.5.1 on an object along one of the orthogonal directions in generally the z-direction or axial direction. Application of the magnetic field causes the protons in that section to orient themselves in the direction of the magnetic field and limits the imaging to this section. The slices that are not under the magnetic field are oriented randomly and hence will not be affected by the subsequent application of the magnetic field or RF pulses.

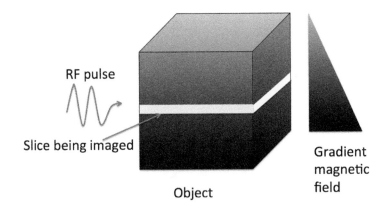

FIGURE 14.9: Slice selection gradient.

14.5.2   Phase Encoding

Phase encoding gradient is generally applied in the x-, y- or z-direction. Due to the application of slice selection gradient, the various protons are oriented in the z-direction (say). They will be spinning in phase with each other. By applying a gradient along the y-direction (say), the protons along a given y-location will spin with the same phase and the other y-locations will spin out of phase. Since every y-location can be identified using its phase, it can be concluded that the protons are encoded with reference to phase. The same argument can be extended if the phase encoding gradient is in the x-direction. If the MRI image has N pixel locations, the phase encoding gradient is chosen such that the phase shift between adjacent pixels is given by Equation 14.5. This ensures that two coordinates do not share the same phase.

$$\phi = \frac{360}{\text{Number of pixels along } x \text{ or } y} \quad (14.5)$$

14.5.3 Frequency Encoding

Frequency encoding gradient is applied in the x-, y- or z- direction. After the application of phase encoding gradient along the y-direction, all the protons along a given y-location will be precessing at the same phase. When a frequency encoding gradient is applied along the x-direction, protons for a given x-location will receive the same magnetic field. Hence these protons will precess at the same frequency. **Thus, with the application of both phase and frequency encoding gradient, every x-, y- point in the object will have a unique phase and frequency.**

14.6 MRI Construction

A simple model (Figures 14.10 and 14.11) of an MRI will consist of:

- Main magnet

- Gradient magnet

- Radio-frequency coils

- Computer for processing the signal

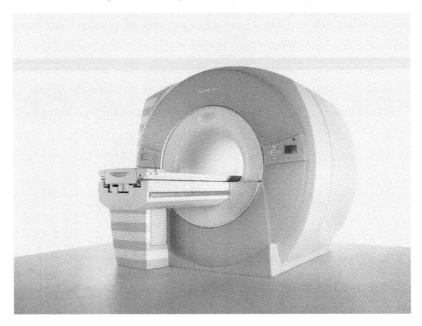

FIGURE 14.10: Closed magnet MRI machine. Original image reprinted with permission from Siemens AG.

14.6.1   Main Magnet

The main magnet generates a strong magnetic field. A typical MRI machine used for medical diagnosis is around 1.5T, which is 30,000 times stronger than the earth's magnetic field.

The magnets could be permanent magnets, electromagnets or super-conducting magnets. An important criterion for choosing a magnet is its ability to produce a uniform magnetic field. Permanent magnets are cheaper but the magnetic field is not uniform. Electromagnets can be manufactured to close tolerance, so that the magnetic field is uniform. They generate a lot of heat, which limits the magnetic field strength. Superconducting magnets are electromagnets that are cooled by super-conducting fluids such as liquid nitrogen or helium. These magnets have a homogeneous magnetic field and high field strength but they are expensive to operate.

FIGURE 14.11: Open magnet MRI machine. Original image reprinted with permission from Siemens AG.

14.6.2 Gradient Magnet

As described earlier, a uniform magnetic field cannot localize the various parts of the object. Hence gradient magnetic fields are used. Based on Faraday's law, a magnetic field can be generated by the application of current to a coil, also known as gradient coils. Since gradient needs to be generated in all three directions, gradient coils are configured to generate fields in all three directions.

14.6.3 RF Coils

An RF coil is composed of loops of conducting materials such as copper. It generates a magnetic field with the passage of current. This process is called transmitting signal. Similarly, a rapidly changing magnetic field generates current in the coil which can be measured. This is accomplished using a receiving coil. In some cases, the same coil can transmit and receive signals. Such coils are called transceivers.

An example of a brain imaging coil is shown in Figure 14.12. Specialized coils are created for different parts being imaged.

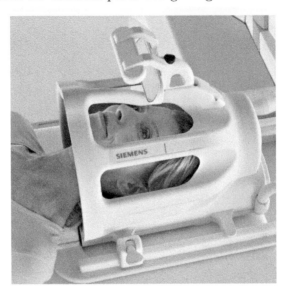

FIGURE 14.12: Head coil. Original image reprinted with permission from Siemens AG.

14.6.4   K-Space Imaging

In Section 14.4, we discussed that the protons regain their orientation after the removal of the RF pulse. During this process, an FID signal (Figure 14.8) is induced in the coil. The FID signal is a plot over time of the change in the net-magnetization in the transverse plane. This signal contains various frequencies that can be obtained using Fourier transformation (Chapter 7). This signal is a 1D signal, as the originating signal is also 1D.

In Section 14.5, we also discussed that the three magnetic field gradients allow signal localization. The three magnetic fields are applied, and the signal obtained for each condition is read out. This 1D signal fills one horizontal line in the frequency space (Figure 14.13). By repeating the signal generation process for all conditions, the various horizontal lines can be filled.

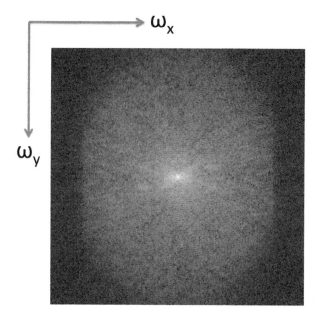

FIGURE 14.13: k-space image.

It can be proven that the image acquired in Figure 14.13 is the Fourier transform of the MRI image. A simple inverse Fourier transform can be used to obtain the MRI image (Figure 14.14). Figure 14.14(a) is the image obtained by filling the k-space and Figure 14.14(b) is obtained using the inverse Fourier transform of the first image.

14.7 T_1, T_2 and Proton Density Image

A typical MRI image consists of T_1, T_2 and proton density weighted components. It is possible to obtain pure T_1, T_2 and a proton density weighted image but it is generally time consuming. Such images are used for discussion to emphasize the role each of these components plays in MRI imaging.

(a) Image obtained by filling the k-space.

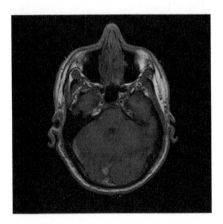

(b) Inverse Fourier transform of k-space image.

FIGURE 14.14: k-space reconstruction of MRI images.

Figure 14.15(a) is a T_1 weighted image (i.e., the pixel values are dependent on the T_1 relaxation time). Similarly, Figure 14.15(b) and Figure 14.15(c) are T_2 and proton density weighted images respectively.

Bright pixels in a T_1 weighted image correspond to fat, while the same pixels appear darker in a T_2 weighted image and vice versa. A proton density image is useful in identifying the pathology of the object.

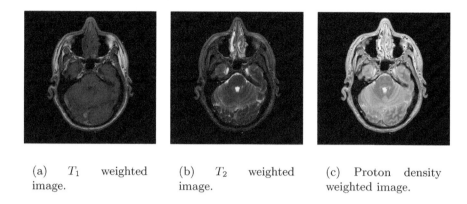

(a) T_1 weighted image.

(b) T_2 weighted image.

(c) Proton density weighted image.

FIGURE 14.15: T_1, T_2 and proton density image. Courtesy of the Visible Human Project.

14.8 MRI Modes or Pulse Sequence

So far, we have learned about the various controls such as gradient magnitude along the three axes and the RF pulse that tilts the orientation of the protons. In this section, we will combine these four controls to produce images that are medically and scientifically useful. This process consists of performing different operations at different times and is generally shown using a pulse sequence diagram. In this diagram, each control receives its own row of operation. The time progresses to the right of each row. Some of these pulse sequences are discussed below. In each case, a certain set of operations or sequences is repeated at regular intervals called the repetition time or TR. TE is defined as the time between the start of the first RF pulse and the time to reach the peak of the echo (or output signal).

14.8.1 Spin Echo Imaging

The spin echo pulse sequence (Figure 14.16) is one of the simplest and most commonly used pulse sequences. It consists of a 90-degree pulse followed by 180-degree pulse at TE/2. During both the pulses, the gradient magnitude along the z-axis is kept on. An echo is produced at time TE while the gradient along the x-axis is kept on so that localization information can be obtained. This process is repeated after every time to repeat (TR). The last 90-degree pulse in the figure is the start of the next sequence.

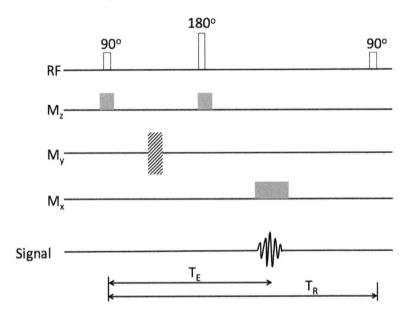

FIGURE 14.16: Spin echo pulse sequence.

14.8.2 Inversion Recovery

The inversion recovery pulse sequence (Figure 14.17) is similar to the spin echo sequence except for a 180-degree pulse applied before the 90-degree pulse.

The 180-degree pulse causes the net-magnetization vector to be inverted along the z-axis. Since the inversion cannot be measured in planes other than xy-plane, a 90-degree pulse is applied. The time between the two pulses is called the inversion time or TI. The gradient magnetic field along the z-axis is kept on during both pulses. The gradient is applied while reading the echo, so that the localization information can be obtained. This process is repeated at regular intervals of TR. The last 180-degree pulse in the figure is the start of the next sequence.

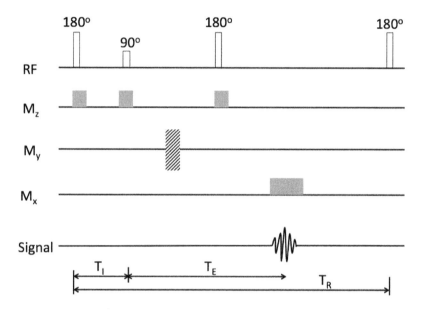

FIGURE 14.17: Inversion recovery pulse sequence.

14.8.3 Gradient Echo Imaging

Gradient echo imaging pulse sequence (Figure 14.18) consists of only one pulse of 90-degree and is one of the simplest pulse sequences. The flip angle could be any angle and 90-degree was chosen as one example. The slice selection gradient is kept on during the application of the 90-degree pulse. At the end of the pulse, a gradient magnetic field is applied along the y-axis. A negative gradient is applied along the x-axis at the same time. The x-axis gradient is then switched to positive gradient while the echo is read. Since there are fewer pulses and the gradient are switched on at consecutive intervals, it is one of the fastest imaging techniques. The last 90-degree pulse in the figure is the start of the next sequence.

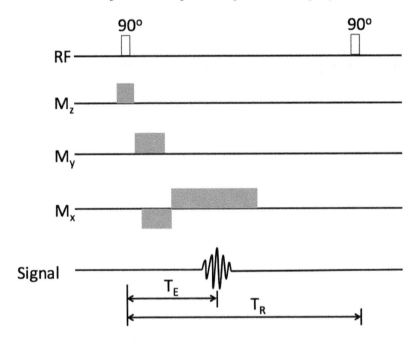

FIGURE 14.18: Gradient echo pulse sequence.

TABLE 14.4: TR and TE settings for various weighted images.

Weighted image	TR	TE
T_1	Short	Short
T_2	Long	Long
Proton density	Long	Short

By using various settings for TR and TE, it is possible to obtain images that are weighted for T_1, T_2 and proton density. A list of such parameters is shown in Table 14.4.

14.9 MRI Artifacts

Image formation in MRI is complex with interaction of various parameters such as homogeneity of the magnetic field, homogeneity of the applied RF signal, shielding of the MRI machine, presence of metal that can alter the magnetic field, etc. Any deviation from ideal conditions will result in an artifact that could either change the shape or the pixel intensity in the image. Some of these artifacts are easily identifiable. For example, a metal artifact leaves identifiable streaks or distortions. A few other artifacts are not easily identifiable. An example of an artifact that is not easily identifiable is the partial volume artifact.

These artifacts can be removed by creating close to ideal conditions. For example, to ensure that there are no metal artifacts, it is important that the patient does not have any implanted metal objects. Alternately, a different imaging modality such as CT or a modified form of MRI imaging can be used.

Artifacts are generally classified into two categories: patient related and machine related. The motion artifact and metal artifact are patient related while inhomogeneity and partial volume artifacts are machine related. An image may contain artifacts from both categories. There are many other artifacts. Interested readers must check the references in Section 14.1.

14.9.1 Motion Artifact

A motion artifact can be due to either motion of the patient or the motion of the organs in a patient. The organ motions occur due to cardiac cycle, blood flow, and breathing. In MRI facilities with poor shielding or design, moving large ferromagnetic objects such as automobiles, elevators, etc., can cause inhomogeneity in the magnetic field that in turn can cause motion artifacts.

Motion due to the cardiac cycle can be controlled by gating, a process of timing the image acquisition with the heart cycle. In some cases, simple breath holding can be used to compensate for the motion artifact.

Figure 14.19 is an example of a slice reconstructed with and without a motion artifact. The motion artifact in Figure 14.19(b) has resulted in significant degradation of the image quality, making clinical diagnosis difficult.

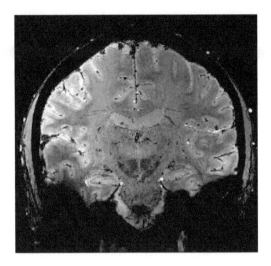

(a) Slice with no motion artifact

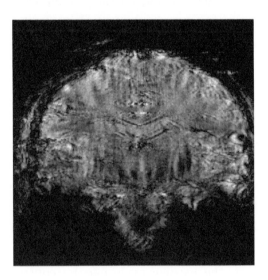

(b) Slice with motion artifact

FIGURE 14.19: Effect of motion artifact on MRI reconstruction. Original images reprinted with permission from Dr. Essa Yacoub, University of Minnesota.

14.9.2 Metal Artifact

Ferromagnetic materials such as iron strongly affect the magnetic field, causing inhomogeneity. In Figure 14.20, the arrows in the two images indicate the direction of the magnetic field. In the left image, the magnetic field surrounds a non-metallic object such as tissue. The presence of the tissue does not change the homogeneity of the magnetic field. In the right image, a metal object is placed in the magnetic field. The magnetic field is distorted close to the object.

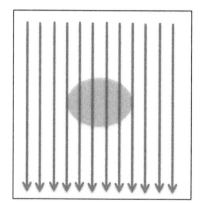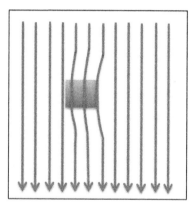

FIGURE 14.20: Metal artifact formation.

The reconstruction process assumes that the field is homogeneous. Thus, it assumes that all points with the same magnetic field strength will have the same Larmor frequency. This variation from ideality causes metal artifact.

The effect is more profound in the case of ferromagnetic materials such as iron, stainless steel, etc. It is less profound in metals such as titanium and other alloys. If MRI is the preferred modality for imaging a patient with metal implants, a low field-strength magnet can be used.

14.9.3 Inhomogeneity Artifact

This artifact is similar in principle to the metal artifact. In the case of metal artifact, the inhomogeneity is caused by the presence of metallic objects. In the case of the inhomogeneity artifact, the magnetic field is not uniform due to a defect in the design or manufacture of the magnet.

The artifact can occur due to the main magnetic field (B_0) or due to gradient magnetic field. In some cases, the main magnetic field may not be uniform across the patient and will change from center to periphery.

The artifact results in distortion depending on the variation of magnetic field across the patient. If the variations are minimal, the shading artifact results.

14.9.4 Partial Volume Artifact

This artifact is caused by imaging using large voxel size, causing two nearby object intensities or pixel intensities to be averaged. This artifact generally affects long and thin objects as their intensity changes rapidly in the direction perpendicular to their long axis.

The artifact can be reduced by increasing the spatial resolution, which results in an increased number of voxels in the image and consequently longer acquisition time.

14.10 Summary

- MRI is a non-radiative high-resolution imaging technique.

- It works on Faraday's law, Larmor frequency, and Bloch equation.

- It is based on physical principles such as T_1 and T_2 relaxation time, proton density, and gyromagnetic ratio.

- Atoms in a magnetic field are aligned in the direction of the magnetic field. An RF pulse can be applied to change their orientation. When the RF pulse is removed, the atoms reorient, and the current generated by this process can be measured. This is the basic principle of NMR.

- In MRI, the basic principle of NMR is used along with slice selection, phase encoding, and frequency encoding gradient to localize the atoms.

- An MRI machine consists of a main magnet, gradient magnets, an RF coil, and a computer for processing.

- The various parameters that control MRI image acquisition are diagrammatically represented as a pulse sequence diagram.

- MRI suffers from various artifacts. These artifacts can be classified as either patient related or machine related.

14.11   Exercises

1. Calculate the Larmor frequency for all atoms listed in Table 14.1.

2. Explain the plot in Figure 14.3 using Equation 14.2.

3. If the plot in Figure 14.3 is viewed looking down in the z direction, the magnetic field path will appear as a circle. Why?

   Solution: The values of M_x and M_y have cos and sin dependencies, similar to the parametric form of a circle.

4. Explain why T_2 is generally smaller than or equal to T_1.

5. Before k-space imaging was used, image reconstruction was achieved using a back-projection technique similar to CT. Write a report about this technique.

6. We discussed a few of the artifacts seen in MRI images. Identify two more artifacts and list their causes, symptoms and method to overcome these artifacts.

7. MRI is generally safe compared to CT. Yet it is important to take precautions during MRI imaging. List some of these precautions.

Chapter 15

Light Microscopes

15.1 Introduction

The modern light microscope was created in the 17th century, but the origin of its important component, the lens, dates back more than three thousand years. The ancient Greeks used lenses as burning glasses, by focusing the sun's rays. In later years, lenses were used to create glasses in Europe in order to correct vision problems. The scientific use of lenses can be dated back to the 16th century with the creation of compound microscopes. Robert Hooke, an English physicist, was the first person to describe cells using a microscope. Antonie van Leeuwenhoek, a Dutch physicist, improved on the lens design and made many important discoveries. For all his research efforts, he is referred to as "the father of microscopy."

We begin this chapter with an introduction to the various physical principles that govern image formation in light microscopy. These include geometric optics, diffraction limit of the resolution, the objective lens, and the numerical aperture. The aim of a microscope is to magnify an object while maintaining good resolving power (i.e., the ability to distinguish two objects that are nearby). The diffraction limit, the objective lens, and the numerical aperture determine the resolving power of a microscope. We apply these principles during the discussion on design of a simple wide-field microscope. This is followed by the fluorescence microscope that not only images the structure but also encodes the functions of the various parts of the specimen. We then discuss confocal and Nipkow disk microscopes that offer better contrast resolution compared to wide-field microscopes. We conclude with a discussion on choosing a wide-field or confocal microscope for a given imaging task. Interested readers can refer to [Bir11], [Dim12],[HL93],[Mer10],[RDLF05],[Spl10] for more details.

15.2   Physical Principles

15.2.1   Geometric Optics

A simple light microscope of today is shown in Figure 15.1. It consists of an eyepiece, an objective lens, the specimen to be viewed and a light source. As the name indicates, the eyepiece is the lens for viewing the sample. The objective is the lens closest to the sample. The eyepiece and the objective lens are typically compound convex lenses. With the introduction of digital technology, the viewer does not necessarily look at the sample through the eyepiece but instead a camera acquires and stores the image.

The lenses used in a microscope have magnification that allows objects to appear larger than their original size. Thus, the magnification for the objective can be defined as the ratio of the height of the image formed to the height of the object. Applying triangular inequality (Figure 15.2), we can also obtain the magnification, m, as the ratio of d_1 to d_0.

$$m = \frac{h_1}{h_0} = \frac{d_1}{d_0} \tag{15.1}$$

A similar magnification factor can be obtained for the eyepiece as well. The total magnification of the microscope can be obtained as the product of the two magnifications.

$$M = m_{\text{objective}} * m_{\text{eyepiece}} \tag{15.2}$$

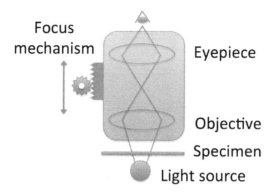

(a) A schematic of the light microscope

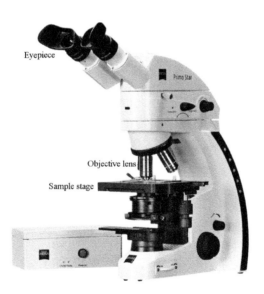

(b) The light microscope. Original image reprinted with permission from Carl Zeiss Microscopy, LLC.

FIGURE 15.1: Light microscope.

15.2.2 Numerical Aperture

Numerical aperture defines both the resolution and the photon-collecting capacity of a lens. It is defined as:

$$NA = n \sin \theta \qquad (15.3)$$

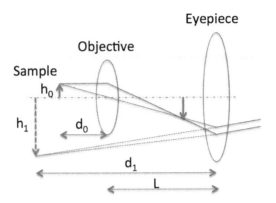

FIGURE 15.2: Schematic of the light microscope.

where θ is the angular aperture or the acceptance angle of the aperture and n is the refractive index.

For high-resolution imaging, it is critical (as will be discussed later) to use an objective with a high numerical aperture. Figure 15.3 is a photograph of an objective with all the parameters embossed. In this example, 20X is the magnification and 0.40 is the numerical aperture.

15.2.3 Diffraction Limit

Resolution is an important characteristic of an imaging system. It defines the smallest detail that can be resolved (or viewed) using an optical system like the microscope. The limiting resolution is called the diffraction limit. We know that electromagnetic radiations have both particle and wave natures. The diffraction limit is due to the wave nature. The Huygens-Fresnel principle suggests that an aperture such as a lens creates secondary wave sources from an incident plane wave. These secondary sources create an interference pattern and produce the Airy disk.

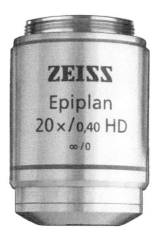

FIGURE 15.3: Markings on the objective lens. Original image reprinted with permission from Carl Zeiss Microscopy, LLC.

Based on the diffraction principles, we can derive the resolving power of a lens. It is the minimum distance between two adjacent points that can be distinguished through a lens. It is defined as:

$$RP = \frac{0.61\lambda}{NA} \qquad (15.4)$$

If a microscope system consists of both objective and eyepiece, then the formula has to be modified to:

$$RP = \frac{1.22\lambda}{(NA_{obj} + NA_{eye})} \qquad (15.5)$$

where NA_{obj} and NA_{eye} are the numerical apertures of the objective and eyepiece respectively.

The aim of any optical imaging system is to improve the resolving power or reduce the value of RP. This can be achieved by decreasing the wavelength, increasing the aperture angle, or increasing the refractive index. Since this discussion is on the optical microscope, we are limited to the visible wavelength of light. X-rays, gamma rays, etc., have shorter wavelengths compared to visible light and hence better resolving power. The refractive index (discussed later) of air is 1.00. The refractive index of mediums used in microscopy imaging is generally greater than 1.00 and hence improves resolving power.

Two points separated by large distances will have distinct Airy disks and hence can be easily identified by an observer. If the points are close (middle image in Figure 15.4), the two Airy disks begin to overlap. If the distance between points is further reduced (left image in Figure 15.4), they begin to further overlap. The two peaks approach and the limit at which the human eye cannot separate the two points is referred to as the Rayleigh Criterion.

FIGURE 15.4: Rayleigh Criterion.

15.2.4   Objective Lens

In the setup shown in Figure 15.1, the two sources of magnification are the objective lens and eyepiece. Since the objective is the closest to the specimen, it is the largest contributor of magnification. Thus, it is critical to understand the inner workings of the objective lens and also the various choices.

We begin the discussion with the refractive index. It is a dimensionless number that describes how electromagnetic radiation passes through various mediums. The refractive index can be seen in various phenomena such as rainbows, separation of visible light by prisms,

TABLE 15.1: List of the commonly used media and their refractive indexes.

Medium	Refractive Index
Air	1.0
Water	1.3
Glycerol	1.44
Immersion oil	1.52

etc. The refractive index of the lens is different from that of the specimen. This difference in refractive index causes the deflection of light. The refractive index between the objective lens and the specimen can be matched by submerging the specimen in a fluid (generally called medium) with the refractive index close to the lens.

Table 15.1 shows commonly used media and their refractive indexes. Failure to match the refractive index will result in loss of signal, contrast and resolving power.

To summarize, the objective lens selection is based on the following parameters:

1. refractive index of the medium,

2. magnification needed,

3. resolution, which in turn is determined by the choice of numerical aperture.

15.2.5 Point Spread Function (PSF)

In Chapter 4, we discussed that Gaussian smoothing is used to reduce noise in an image. The noise reduction is achieved by smearing the pixel value at one location to all its neighbors. Any optical system performs a similar operation with a kernel called a Point Spread Function (PSF). It is the response of an optical system to a point input or point object as a consequence of diffraction. When a point source of light is passed through a pinhole aperture, the resultant image on a focal plane is not a point, but instead the intensity is spread to multiple neighboring pixels. In other words, the point image is blurred by the PSF. The PSF is dependent on the numerical aperture of the lens. A lens with a high numerical aperture produces a PSF of smaller width.

15.2.6   Wide-Field Microscopes

Light microscopes can be classified into different types depending on the method used to acquire images, improve contrast, illuminate samples etc. The microscope that we have described is called a wide-field microscope. It suffers from poor spatial resolution (without any computer processing), and poor contrast resolution due to the effect of PSF.

15.3   Construction of a Wide-Field Microscope

A light microscope (Figure 15.1) is designed to magnify the image of a sample using multiple lenses. It consists of the following:

1. Eyepiece

2. Objective

3. Light source

4. Condenser lens

5. Specimen stage

6. Focus knobs

The eyepiece is the lens closest to the eye. Modern versions of the eyepiece are compound lenses in order to compensate for aberrations.

It is interchangeable and can be replaced by eyepieces of different magnification depending on the nature of object being imaged.

The objective is the lens closest to the object. These are generally compound lenses in order to compensate for aberrations. They are characterized by three parameters: magnification, numerical aperture and the refractive index of the immersion medium. The objectives are interchangeable and hence modern microscopes also contain a turret that contains multiple objectives to enable easier and faster switching between different lenses. The objective might be immersed in oil to match the refractive index and increase the numerical aperture and consequently increase the resolving power.

The light source is at the bottom of the microscope. It can be tuned to adjust the brightness in the image. If the lighting is poor, the contrast of the resultant image will be poor, while excess light might saturate the camera recording the image. The most commonly used illumination method is Köhler illumination, designed by August Köhler in 1893. The previous methods suffered from non-uniform illumination, projection of the light source on the imaging plane, etc. Köhler illumination eliminates non-uniform illumination so that all parts of the light source contribute to specimen illumination. It works by ensuring that the lamp image is not projected on the sample plane with the use of a collector lens placed near the lamp. This lens focuses the image of the lamp to the condenser lens. Under this condition, illumination of the specimen is uniform.

The specimen stage is used for placing the specimen under examination. The stage can be adjusted to move along its two axes, so that a large specimen can be imaged. Depending on the features of a microscope, the stage could be manual or motor controlled.

Focus knobs allow moving the stage or objective in the vertical axis. This allows focusing of the specimen and also enables imaging of large objects.

15.4   Epi-Illumination

In the microscope setup shown in Figure 15.1, the specimen is illuminated by using a light source placed below. This is called transillumination. This method does not separate the emission and excitation light in fluorescence microscopy. An alternate method called epi-illumination is used in modern microscopes.

In this method (Figure 15.7), the light source is placed above the specimen. The dichroic mirror reflects the excitation light and illuminates the specimen. The emitted light (which is of longer wavelength) travels through the dichroic mirror and is either viewed or detected using a camera. Since there are two clearly defined paths for emission and excitation light, only the emitted light is used in the formation of the image and hence improves the quality of the image.

15.5   Fluorescence Microscope

A fluorescence microscope allows identification of various parts of the specimen not only in terms of structure but also in terms of function. It allows tagging different parts of the specimen, so that it generates light of certain wavelengths and forms an image. This improves the contrast in the image between various objects in a specimen.

15.5.1   Theory

Fluorescence is generally observed when a fluorescent molecule absorbs light at a particular wavelength and emits light at a different wavelength within a short interval. The molecule is generally referred to as fluorochrome or dye, and the delay between absorption and emission is in the order of nanoseconds. This process is generally shown diagrammatically using the Jablonski diagram shown in Figure 15.5. The figure should be read from bottom to top. The lower state is the stable ground state, generally called the S_0 state. A light or photon incident on the fluorochrome causes the molecule to reach an excited state (S'_1). A molecule in the excited state is not stable and hence returns to its stable state after losing the energy both in the form of radiation such as heat and also light of longer wavelength. This light is referred to as the emitted light.

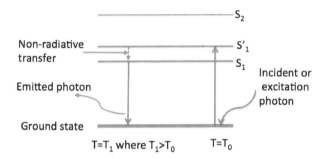

FIGURE 15.5: Jablonski diagram.

From Planck's law that was discussed in Chapter 13, the energy of light is inversely proportional to the wavelength. Thus, light of higher energy will have shorter wavelength and vice versa. The incident photon is a higher energy and hence shorter wavelength, while the emitted light is of low energy and longer wavelength. The exact mechanisms of emission and absorption are beyond the scope of this book and readers are advised to consult books dedicated to fluorescence for details.

15.5.2 Properties of Fluorochromes

Two properties of fluorochromes, excitation wavelength and emission wavelength, were discussed in the previous section. Table 15.2 lists the excitation and emission wavelengths of commonly used fluorochromes. As can be seen in the table, the difference between excitation and emission wavelengths, or the Stokes shift, is significantly

TABLE 15.2: List of the fluorophores of interest to fluorescence imaging.

Fluorochrome	Peak Excitation Wavelength (nm)	Peak Emission Wavelength (nm)	Stokes Shift (nm)
DAPI	358	460	102
FITC	490	520	30
Alexa Fluor 647	650	670	20
Lucifer Yellow VS	430	536	106

different for different dyes. The larger the difference, the easier it is to filter the signal between emission and excitation.

A third property, quantum yield, is another important characteristic of a dye. It is defined as:

$$QY = \frac{\text{Number of emitted photons}}{\text{Number of absorbed photons}} \qquad (15.6)$$

Another important property that determines the amount of fluorescence generated is the absorption cross-section. The absorption cross-section can be explained with the following analogy. If a bullet is fired at a target, the ability to reach the target is better if the target is large and if the target surface is oriented in the direction perpendicular to the direction of the bullet path. Similarly, the term absorption cross-section defines the "effective" cross-section of the fluorophore and hence the ability of the excitation light to produce fluorescence.

It is measured by exciting a sample of fluorophore of certain thickness with excitation photons of a certain intensity and measuring the intensity of the emitted light. The relationship between the two intensities is given by Equation 15.7.

$$I = I_0 e^{-\sigma D \delta x} \qquad (15.7)$$

where I_0 is the excitation photon intensity, I is the emitted photon intensity, σ is the absorption cross-section of the fluorophore, D is the density, and δx is the thickness of the fluorophore.

15.5.3 Filters

During fluorescence imaging, it is necessary to block all light that is not emitted by the fluorochrome to ensure the best contrast in the image and consequently better detection and image processing. In addition, the specimen does not necessarily contain only one type of fluorochrome. Thus, to separate the image created by one fluorochrome from the other, a filter that allows only light of a certain wavelength corresponding to the different fluorochromes is needed.

The filters can be classified into three categories: lowpass, bandpass and highpass. We discussed these as digital filters in Chapter 7 while here we will discuss physical filters. The lowpass filter allows light of shorter wavelength and blocks longer wavelengths. The highpass filter allows light of longer wavelength and blocks shorter wavelengths. The bandpass filter allows light of a certain range of wavelengths. In addition, fluorescence microscopy uses a special type of filter called a dichroic mirror (Figure 15.7). Unlike the three filters discussed earlier, in a dichroic mirror the incident light is at 45° to the filter. The mirror reflects light of shorter wavelength and allows longer wavelengths to pass through.

Multi-channel imaging is a mode where different types of fluorochromes are used for imaging resulting in images with multiple channels. Such images are called multi-channel images. Each channel contains an image corresponding to one fluorochrome. For example, if we obtained an image of size 512-by-512, using two different fluorochromes, the image would be of size 512-by-512-by-2. The two in the size corresponds to the two channels. Generally most fluorescence images have 3 dimensions. Hence the volume in such cases would be 512-by-512-by-200-by-2, where 200 is the number of slices or z-dimension. The actual number may vary based on the imaging conditions.

The choice of the fluorochrome is dependent on the following parameters:

1. Excitation wavelength

2. Emission wavelength

3. Quantum yield

4. Photostability

Filters used in the microscope need to be chosen based on the fluorochrome being imaged.

15.6　Confocal Microscopes

Confocal microscopes overcome the issue of spatial resolution that affects wide-field microscopes. A better resolution in confocal microscopes is achieved by the following:

- A narrow beam of light illuminates a region of the specimen. This eliminates collection of light by the reflection or fluorescence due to a nearby region in the specimen.

- The emitted or reflected light arising from the specimen passes through a narrow aperture. A light emanating from the direction of the beam will pass through the aperture. Any light emanating from nearby objects or any scattered light from various objects in the specimen will not pass through the aperture. This process eliminates all out-of-focus light and collects only light in the focal plane.

The above process describes image formation at a single pixel. Since an image of the complete specimen needs to be formed, the narrow beam of light needs to be scanned all across the specimen and the emitted or reflected light needs to be collected to form a complete image. The scanning process is similar to the raster scanning process used in television. It can be operated using two methods. In the first method devised by Marvin Minsky, the specimen is translated so that all points can be scanned. This method is slow and also changes the shape of the specimen suspended in liquids and is no longer used. The second approach is to keep the specimen stationary while the light beam is scanned across the specimen. This was made possible by advances in optics and computer hardware and software, and is used in all modern microscopes.

15.7 Nipkow Disk Microscopes

Paul Nipkow created and patented a method for converting an image into an electrical signal in 1884. The method consisted of scanning an image by using a spinning wheel containing holes placed in a spiral pattern, as shown in Figure 15.6. The portion of the wheel that does not contain the hole is darkened so that light does not pass through it. By rotating the disk at constant speed, a light passing through the hole scans all points in the specimen. This approach was later adapted to microscopy. Figure 15.6 shows only one spiral with a smaller number of holes while a commercially available disc will have a large number of holes, to allow fast image acquisition.

A setup containing the disk along with the laser source, objective lens, detector, and the specimen is shown in Figure 15.7, and Figure 15.8 is a photograph of a Nipkow disk microscope. In this figure, the illuminating light floods a significant portion of the holes. The portion that does not contain any holes reflects the light. The light that passes through the holes reaches the specimen through the objective lens. The reflected light, or the light emitted by fluorescence, passes through the

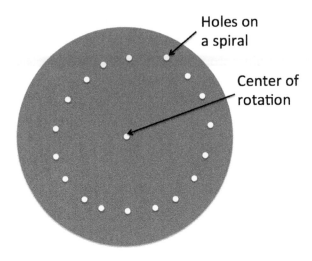

FIGURE 15.6: Nipkow disk design.

objective and is reflected by the dichroic mirror. The detector forms an image using the reflected light.

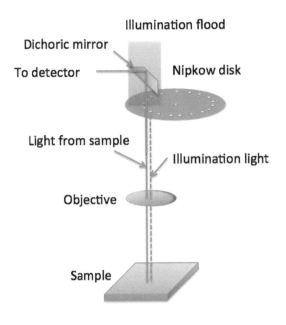

FIGURE 15.7: Nipkow disk setup.

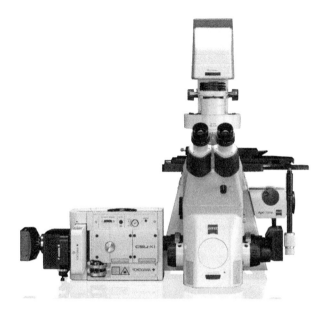

FIGURE 15.8: Photograph of Nipkow disk microscope. Original image reprinted with permission from Carl Zeiss Microscopy, LLC.

Unlike a regular confocal microscope, the Nipkow disk microscope is faster as neither the specimen nor the light beam needs to be raster scanned. This enables rapid imaging of live cells.

15.8 Confocal or Wide-Field?

Confocal and wide-field microscopes each have their own advantages and disadvantages. These factors need to be considered when making a decision on what microscope to use for a given cost or type of specimen or the analysis to be performed.

- Resolution: There are two different resolutions: xy and z directions. Confocal microscopes produce better resolution images in both directions. Due to advances in computing and better software, wide-field images can be deconvolved [WSS01] to a good resolution along x and y but not necessarily along the z direction.

- Photo bleaching: Images from a confocal microscope may be photo-bleached, as the specimen is imaged over a longer time period compared to a wide-field microscope.

- Noise: Wide-field microscopes generally produce images with less noise due to blurring from the PSF.

- Acquisition rate: Since confocal images scan individual points, it is generally slower compared to wide-field microscopes.

- Cost: As a wide-field microscope has fewer parts, it is less expensive than confocal.

- Computer processing: Confocal images need not be processed using deconvolution. Depending on the setup, deconvolution of a wide-field image can produce images of comparable quality to confocal images.

- Specimen composition: A wide-field microscope with deconvolution works well for a specimen with a small structure.

15.9   Summary

- The physical properties that govern optical microscope imaging are magnification, diffraction limits, and numerical aperture. The diffraction limit and numerical aperture determine the resolution of the image.

- The specimen is immersed in a medium in order to match the refractive index and to increase the resolution.

- Wide-field and confocal are the two most commonly used microscopes. In the former, a flood of light is used to illuminate the specimen while in the latter, a pencil beam is used to scan the specimen and the collected light passes through a confocal aperture.

- The fluorescence microscope allows imaging of the shape and function of the specimen. Fluorescence microscope images are obtained after the specimen has been treated with a fluorophore.

- The specific range of wavelength emitted by the fluorophore is measured by passing the light through a filter.

- To speed up confocal image acquisition, a Nipkow disk is used. The disk consists of a series of holes placed on a spiral. The disk is rotated and the position of the holes is designed to ensure that complete 2D scanning of the specimen is achievable.

15.10 Exercises

1. If the objective has a magnification of 20X and the eyepiece has a magnification of 10X, what is the total magnification?

2. A turret has three objectives: 20X, 40X and 50X. The eyepiece has magnification of 10X. What is the highest magnification achievable?

3. In the same turret setup, if a cell occupies 10% of the field of view for an objective magnification of 20X, what would be the field of view percentage for 40X?

4. Discuss a few methods to increase spatial resolution in an optical microscope. What are the limits for each parameter?

Chapter 16

Electron Microscopes

16.1 Introduction

The resolution of a light microscope discussed previously is directly proportional to the wavelength. To improve the resolution, light with a shorter wavelength should be used. Scientists began experimenting with ultraviolet light, which has a shorter wavelength than visible light. Due to the difficulty in generation and maintaining coherence, it was not commercially successful.

Meanwhile, the French physicist Louis de Broglie proved a traveling electron has both wave and particle duality similar to light. He was awarded a Nobel Prize in 1929 for this work.

An electron wave with higher energy will have lower wavelength and vice versa. Thus, improving the resolution would involve increasing the energy. The wavelength of electrons is considerably shorter than that of visible light and hence very high-resolution images can be obtained. Visible light has a wavelength of 400–700 nm. Electrons, on the other hand, have a wavelength of 0.0122 nm for an accelerating voltage of 10 kV.

Ernst Ruska and Max Knoll created the first electron microscope (EM) with the ability to magnify objects 400 times. Upon further work, Ruska improved its resolution beyond the resolution of optical microscopes and hence made the EM an indispensable tool for microscopists. The EM used today does not measure a single characteristic property, but rather measures multiple characteristics of the material. The one common thing among all of them is the electron beam.

In Section 16.2, we discuss some of the physical principles that need to be understood regarding the EM. We begin with a discussion of the properties of the electron beam and its ability to produce images at high resolution. We introduce the interaction of electrons with the matter and various particles and waves that are generated as a consequence. The fast-moving electron beam from the electron gun passes through the material to be imaged. During its transit through the material, the electron interacts with the atoms in that material. We integrate the two basic principles and discuss the construction of an EM. We also discuss specimen preparation and general precautions when preparing the specimen. Interested readers can refer to [BR98],[DR03],[HK93],[Gol03],[Haj99],[Hay00],[Key97], [KB46],[Kuo07],[Sch89],[Spl10],[Wat97].

16.2   Physical Principles

The EM was made possible by many fundamental and practical discoveries made over time. In this section, we discuss these discoveries and place them in the context of creating an electron microscope.

EM process involves bombarding a specimen with a high-speed electron beam and recording the beam emanating from or transmitted through the specimen. These high-speed electrons have to be focused to a point in the specimen. In 1927, Hans Busch proved that an electron beam can be focused on an inhomogeneous magnetic field just as light can be focused using a lens. Four years later, Ernst Ruska and Max Knoll confirmed this by constructing such a magnetic lens. This lens is still a part of today's EM design.

The second basic principle is the dual nature of the electron beam proven by Louis de Broglie. The electron beam behaves as a wave and a particle just like visible light. Thus the electron beam has both wavelength and mass.

16.2.1 Electron Beam

Louis de Broglie proved that electrons traveling at high speed have both particle and wave characteristics. The wavelength of the beam is given by Equation 16.4. Thus the faster the electrons travel, the lower the wavelength of the beam. As we will discuss later, the lower wavelength results in production of high-resolution images.

$$\lambda = \frac{h}{mv} \tag{16.1}$$

where h is Planck's constant and is equal to $6.626 \ 10^{-34}$ Js, m is the electron mass and is equal to $m = 9.109 \ 10^{-31}$ kg, and v is the velocity of the particle.

We also know that the beam stores the energy in the form of kinetic energy given by the following equation.

$$E = \frac{mv^2}{2} = eV \tag{16.2}$$

where $e = 1.602 \ 10^{-19}$ coulombs is the charge of the electron and V is the acceleration voltage. In other words,

$$v = \sqrt{\frac{2eV}{m}} \tag{16.3}$$

Plug this equation into Equation 16.1 to obtain

$$\lambda = \frac{h}{\sqrt{2meV}} \tag{16.4}$$

Since all the variables on the right-hand side of the equation are constant except for the accelerating voltage V, we can simplify the equation to

$$\lambda = \frac{1.22}{\sqrt{V}} \qquad \text{nano-meter} \tag{16.5}$$

where V is voltage measured in volts. Thus for an accelerating voltage of 10 kV, the wavelength of the electron beam is 0.0122 nm.

Since the speed of the beam and the acceleration voltage are generally very high for electron microscopy, the wavelength computation is dependent on the relativistic effect. For such cases, it can be shown that the wavelength is

$$\lambda = \frac{h}{\sqrt{2mE\nu\left(1 + \frac{eV}{\frac{2m}{c^2}}\right)}} \tag{16.6}$$

16.2.2 Interaction of Electron with Matter

In Chapter 13, "X-Ray and Computed Tomography" we discussed the interaction of x-rays with materials. We discussed the Bremsstrahlung spectrum (braking spectrum) and the characteristic spectrum. The former is created as the incident x-ray is slowed by its passage through the material. The latter is formed when the x-rays knock out electrons from their orbit.

The electron beam has both particle and wave natures similar to x-rays. Hence the electron beam exhibits a spectrum similar to x-rays. Since the energy of the electron is higher than that of the x-ray, it also produces few other emissions. The various emissions are:

1. transmitted electrons,

2. back-scattered electrons (BSEs),

3. secondary electrons (SEs),

4. elastically scattered electrons,

5. inelastically scattered electrons,

6. Auger electrons (AEs),

7. characteristic x-rays,

8. Bremsstrahlung x-rays,

9. visible light (cathodoluminesence),

10. diffracted electrons (DEs), and

11. heat.

These phenomena are shown in Figure 16.1. The various emissions occur at different depths of the material. The region that generates these emissions is referred to as the electron interaction volume. SEs are generated at the top of the region while the Bremsstrahlung x-rays are generated at the bottom.

In a typical EM, not all of these are measured. For example, in the transmission EM or TEM, the transmitted electron, elastically scattered electron and inelastically scattered electrons are measured, and in the scanning EM (SEM), BSEs or SEs are measured.

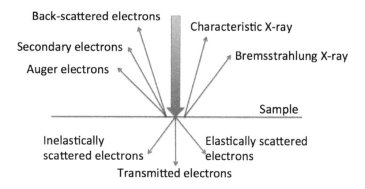

FIGURE 16.1: Intensity distributions.

Since we discussed Bremsstrahlung and characteristic x-rays earlier, we will focus on the other important emissions, the BSEs, SEs and TEs, in this chapter.

16.2.3   Interaction of Electrons in TEM

The TEM measures three different electrons during its imaging process. They are transmitted electrons, elastically scattered (or diffracted electrons), and inelastically scattered electrons.

In Chapter 13, we discussed image formation by exposing a photographic plate or a digital detector to an x-ray beam after it passes through material. The image is formed using the varying intensity of the x-ray in proportion to the thickness and attenuation coefficient of the material at various points. In the TEM, the incident beam of electrons replaces the x-ray. This beam is transmitted through the specimen without any significant change in intensity, unlike x-ray. This is due to the fact that the electron beam has very high energy and that the specimen is extremely thin (on the order of 100 microns). The region in the specimen that is opaque will transmit fewer electrons and appear darker.

A part of the beam is scattered elastically (i.e., with no loss of energy) by the atoms in the specimen. These electrons follow Bragg's law of diffraction. The resultant image is a diffraction pattern.

The inelastically scattered electrons (i.e., with loss of energy) contribute to the background. The specimen used in the TEM is generally very thin. Increasing the thickness of the specimen results in more inelastic scattering and hence more background.

16.2.4   Interaction of Electrons in SEM

The TEM specimens are generally thin and hence there are fewer modes of interaction. The SEM, on the other hand, uses a thick or bulk specimen and hence has more modes of interaction in addition to the modes discussed in Section 16.2.3.

In the SEM, the various modes of interaction are:

1. Characteristic x-rays,

2. Bremsstrahlung x-rays,

3. Back-scattered electrons (BSEs),

4. Secondary electrons (SEs),

5. Auger electrons

6. Visible light, and

7. Heat.

The generation of characteristic x-rays and Bremsstrahlung x-rays was discussed in Chapter 13. The former is produced by the knock-out of an electron from its orbit by the fast-moving electron, while the latter is produced by deceleration of the electron during its transit through material.

The mechanism of generation of Auger electrons is similar to characteristic x-rays. When a fast-moving electron ejects an electron in orbit, it leaves a vacancy in the inner shell. An electron from a higher shell fills this vacancy. The excess energy is released as an x-ray in the case of the characteristic x-ray, while an electron is ejected during Auger electron formation. Since the Auger electron has low energy, it is generally formed only on the surface of the specimen.

SEs are low-voltage electrons. They are generally less than 50 eV in energy. They are generally emitted at the top of the specimen, as their energy is too small to be emitted inside the material and still escape to be detected. Since SEs are emitted from the top of the surface, they are used for imaging the topography of the sample.

BSEs are obtained by the scattering of the primary electron by the specimen. This scattering occurs at depths higher than the regions where SEs are generated. Materials with high atomic numbers produce a significantly larger number of BSEs and hence appear brighter in the BSEs detector image. Since BSE are emitted from the inside of the specimen, they are used for imaging the chemical composition of the specimen and also for topographic imaging.

16.3　Construction of EMs

16.3.1　Electron Gun

The electron gun generates an accelerated beam of electrons. There are two different types of electron gun: the thermionic gun and field emission gun. In the former, electrons are emitted by heating a filament while in the latter, electrons are emitted by the application of an extraction potential.

A schematic of the thermionic gun is shown in Figure 16.2. The filament is heated by passing current, which generates electrons by the process of thermionic emission. It is defined as emission of electrons by absorption of thermal energy. The number of electrons produced is proportional to the current through the filament. The Wehnelt cap is maintained at a small negative potential, so that the negatively charged electrons are accelerated in the direction shown through the small opening. The anode is maintained at a positive potential, so that the electrons travel down the column toward the specimen. The acceleration is achieved by the voltage between the cap and the anode.

The filament can be made of tungsten or lanthanum hexaboride (LaB_6) crystals. Tungsten filaments can work at high temperatures but do not produce circular spots. The LaB_6 crystals on the other hand, can produce circular spots and hence better spatial resolution.

A schematic of the field emission gun (FEG) is shown in Figure 16.3. The filament used is a sharp tungsten metal tip. The tip is sharpened to have a dimension on the order of 100 nm. In a cold FEG, the electron from the outer shell is extracted by using the extraction voltage (V_E). The extracted electrons are accelerated using the accelerating voltage (V_A). In the thermionic FEG, the filament is heated to generate electrons. The extracted electrons are accelerated to high energy.

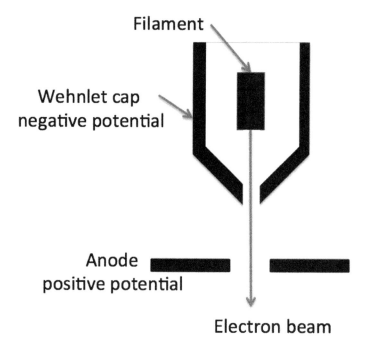

FIGURE 16.2: Thermionic gun.

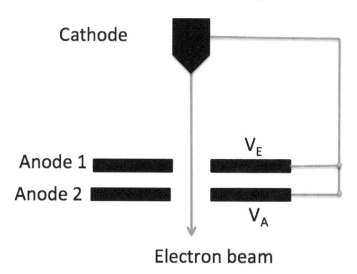

FIGURE 16.3: Field emission gun.

16.3.2   Electromagnetic Lens

In Chapter 15, "Light Microscopes," we discussed the purpose of the various lenses, objective and eyepiece. The lens is chosen such that the light from the object can be focused to form an image. Since electrons behave like waves, they can be focused by using lenses.

From our discussion of the Image Intensifier (II) in Chapter 13, the electrons are affected by the magnetic field. In the case of the II, this phenomenon presents a problem and results in distortion. However, a controlled magnetic field can be used to navigate electrons and hence create a lens. It has been proven that an electron traveling through vacuum in a magnetic field will follow a helical path.

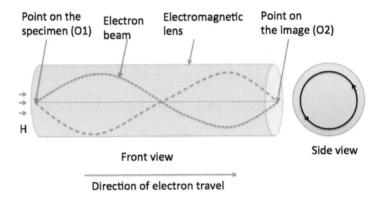

FIGURE 16.4: Electromagnetic lens.

The electrons enter the magnetic field at point O1 (Figure 16.4). Point O2 is the point where all electrons generated by the electron gun are focused by the magnetic field. The distance O1-O2 is the focal length of the lens. The mathematical relationship that defines focal length is given by

$$f = K \frac{V}{i^2} \qquad (16.7)$$

where K is a constant based on the design of the coil at the geometry, V is the accelerating voltage, and i is the current through the coil. As can be seen, either increasing the voltage or reducing the current in the coil can increase focal length. In an optical microscope, the focal length for a given lens is fixed while it can be changed in the case of an electromagnetic lens. Hence, in an optical microscope, the only method for changing the focal length is by either changing the lens (using objective turret) or by changing the spacing between the lenses. On the other hand, in an electromagnetic lens, the magnification can be changed by altering the voltage and current. The electromagnetic lens suffers from aberrations similar to optical lenses. Some of these are astigmatism, chromatic aberration, and spherical aberration. They can be overcome by designing and manufacturing under high tolerance.

16.3.3 Detectors

Secondary electron detectors: SEs are measured using the Everhart-Thornley detector (Figure 16.5). It consists of a Faraday cage, a scintillator, a light guide and a photo-multiplier tube. SEs have very low-energy (less than 50 eV). To attract these low energy electrons, a positive voltage on the order of 100 V is applied to the Faraday cage in order to attract SEs. The scintillator is maintained at a very high positive voltage to attract the SEs. The SEs are converted to light photons by the scintillator. The light generated is too weak to form an image. Hence the light is guided through a light guide onto a photo-multiplier tube, which amplifies the light signal to form an image.

Back-scattered electron detectors: BSEs have very high energy and hence readily travel to a detector. BSEs also travel in all directions and hence a directional detector, such as the Everhart-Thornley detector can only collect a few electrons and will not be enough to form a complete image. BSE detectors are generally doughnut shaped (Figure 16.6) and placed around the electron column just below the objective lens. The detecting element is either a semiconductor or a scintillator that converts the incident electron into light photons that are recorded using a camera.

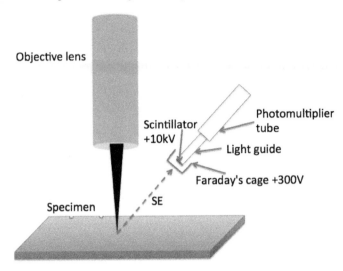

FIGURE 16.5: Everhart-Thornley secondary electron detector.

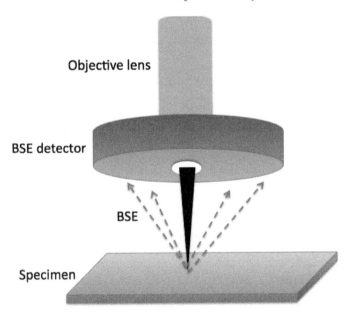

FIGURE 16.6: Back-scattered electron detector.

16.4   Specimen Preparations

The specimen needs to be electrically conductive. Hence, biological specimens are coated with a thin layer of electrically conductive material such as gold, platinum, tungsten, etc. In some cases, a biological sample cannot be coated without affecting the integrity of the specimen. In such cases, an SEM can be operated at low voltage. Materials made of metal do not have to be coated, as they are electrically conductive.

Since the electron beam can only travel in vacuum, the specimen needs to be prepared for placement in a vacuum chamber. Materials with water need to be dehydrated. The dehydration process causes the specimen to shrink and change in shape. Hence the specimen has to be chemically fixed, in which water is replaced by organic compounds. The specimen is then coated with electrically conductive material before imaging.

An alternate method is to freeze the sample using cryofixation. In this method, the specimen is cooled rapidly by plunging into liquid nitrogen (boiling point $= -195.8^oC$). The rapid cooling of the specimen preserves its internal structure so that it can be imaged accurately. The rapid cooling ensures that ice crystals, which can damage the specimen, do not form. In the case of the TEM, since the specimen has to be thin, the cryofixated specimen is cut into thin slices or microtomy.

16.5   Construction of the TEM

In the previous sections, we have discussed the various components of the TEM and SEM. In the next two sections, we will integrate the various parts to construct the TEM and SEM. Figure 16.7 illustrates the bare-bones optical microscope, TEM, and SEM. Although this discussion is for illustration purposes, the complete equipment consists of multiple controls to ensure good image quality.

In each case, a source of light or electrons is at the top. The light in the case of the optical microscope travels through a condenser lens, a specimen, and then the objective or eyepiece, to be either viewed by an eye or imaged using a detector.

In the case of the TEM, the source is the electron gun. The accelerated electrons are focused using a condenser lens, transmitted through the specimen and finally focused to form an image using objective and eyepiece magnetic lenses. Since the electron beam can only travel in vacuum, the entire setup is placed in a vacuum chamber.

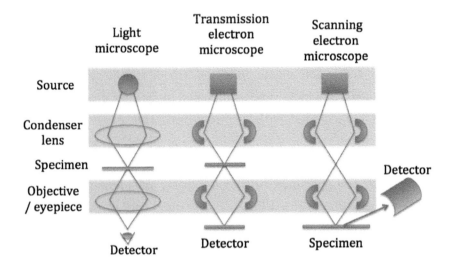

FIGURE 16.7: Comparison of the optical microscope, TEM, and SEM.

An example of an image of Sindbis virus obtained using TEM is shown in Figure 16.8 [ZMP+02].

16.6 Construction of the SEM

Figure 16.7 illustrates the schematics of the light microscope, TEM, and SEM. Figure 16.9 is an example of an SEM machine. In the case of the SEM, the source is the electron gun as in the TEM. The accelerated electron is then focused using a condenser lens to form a small spot on the specimen. The electron beam interacts with the specimen and emits BSEs, SEs, Auger electrons, etc. These are measured using the detector

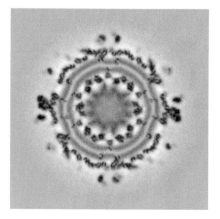 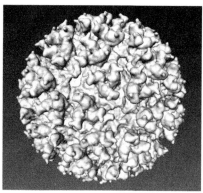

(a) A slice of 3D image obtained using a TEM.

(b) All slices rendered to an iso-surface.

FIGURE 16.8: TEM slice and its iso-surface rendering. Original image reprinted with permission from Dr. Wei Zhang, University of Minnesota.

discussed previously to form an image. Since the electron beam can only travel in vacuum, the entire setup is placed in a vacuum chamber.

An example of an image obtained using an SEM is shown in Figure 16.10.

16.7   Factors Determining Image Quality

The three factors that determine the quality of image are

- voltage,

- working distance,

- spot size.

Voltage: The voltage is generally less than 30 kV. Better contrast (i.e., higher contrast resolution) can be obtained using higher voltage.

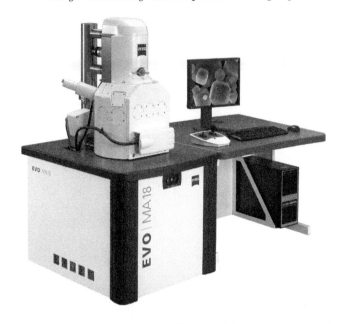

FIGURE 16.9: An SEM machine. Original image reprinted with permission from Carl Zeiss Microscopy, LLC.

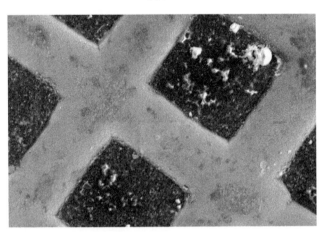

FIGURE 16.10: BSE image obtained using an SEM.

Lower voltage can be used to image a biological specimen without the need for fixation. As discussed earlier, higher voltage produces electrons that can penetrate the specimen deeper and hence larger interaction volume. As the interaction volume increases, the composition of the specimen response to the electron beam changes. For example, at low voltage, auger electrons and secondary electrons are primarily emitted from the interaction of the electron beam and the specimen. At high voltage, back-scattered electrons and x-rays are emitted.

Working distance (WD): It is the distance between the end of the electron column and the top of the specimen. Shorter working distance is used for high-resolution imaging. For flat objects such as semiconductor wafers, WD is almost a constant, while it could vary significantly while imaging biological specimens.

Spot size: The spot size determines the spatial resolution of the image. Smaller spot size results in higher spatial resolution and vice versa.

16.8   Summary

- The EM involves bombarding high-speed electron beams on a specimen and recording its response.

- Imaging an electron is possible, as it exhibits both particle and wave natures.

- The wavelength of the electron is inversely proportional to the square root of the accelerating voltage. Increasing the accelerating voltage results in lower wavelength or higher resolution. The typical accelerating voltage is 30kV.

- A high-speed electron beam bombards a specimen and generates characteristic x-rays, Bremsstrahlung x-rays, back-scattered electrons (BSEs), secondary electrons (SEs), Auger electrons, visible light, and heat. BSEs and SEs are the most commonly measured in the SEM.

- The EM focuses the beam using an electromagnetic lens.

- The BSE is measured using a doughnut-shaped detector wrapped around the axis of the electron beam.

- The SE is measured using an Everhart-Thornley detector.

- Unlike a light microscope, in the case of electron microscopes, the specimen needs to be carefully prepared as the imaging is conducted in vacuum.

- The parameters that determine the quality of image are voltage, working distance, and spot size.

16.9 Exercises

1. The accelerating voltage of an SEM is 10kV. Calculate the wavelength of the generated electron.

2. Compare and contrast the working principles of the TEM and SEM.

3. List the order of generation of various spectrums in the electron interaction volume beginning with the surface of the specimen.

Appendix A

Process-Based Parallelism using Joblib

A.1 Introduction to Process-Based Parallelism

The execution of a Python code launches a Python process that accepts the instruction and executes it. If we need to process multiple images (for example) using the same instruction, then the Python process runs the instruction on each of the images sequentially. If we have 8 images and if each image takes a minute to process, the total processing time would be 8 minutes. However, a typical modern computer has multiple cores each of which can handle one Python process. So it would be preferable to process each of these images in parallel and use all the cores in a modern computer. This can be achieved using Python's joblib module. If a computer has 8 cores and we start 8 Python processes, then the processing described previously can be completed in 1 minute at a speedup of 8X. Modern servers have 12+ cores and hence we can obtain considerable speedup using joblib.

A.2 Introduction to Joblib

The module joblib ([Job20]) is designed to perform process-based parallelism. It has other functionalities but we will limit the discussion only to parallelism. The joblib mechanism for parallelization is a single class called Parallel. This simple mechanism enables easy conversion of an existing serial code to parallel without significant cost to the programmer.

The Parallel class instance takes a generator expression [PC18]. The generator in Python returns an object (also called an iterator) which we can iterate over and fetch one value at a time. The function that needs to be parallelized has to be decorated by joblib's 'delayed' decorator.

We will discuss a few examples where we will compute the value of cube of numbers between 0 and 9 to demonstrate the syntax of a joblib parallel code. We will then complete the task by processing images in parallel using joblib.

A.3   Parallel Examples

In the next three examples, we will parallelize the same functionality. The task to parallelize is defined in the function called cube. The function takes an argument 'x' and returns its cube. In all cases, we import the class Parallel and the decorator delayed from joblib. The parameter n_jobs determines the number of parallel processes. A value of −1 indicates that the number of parallel processes will be equal to the number of cores. If a value of 1 is used, then the number of parallel processes will be 1.

When running the examples below, we recommend opening the process monitor in your operating system, such as Task Manager for Windows, Activity Monitor for Mac, or top on Linux, and notice that new Python processes are created proportional to the value of n_jobs.

In the first example, the cube function is decorated with the delayed decorator. The generator expression fetches each of the values 0, 1 ... 9 and passes it to the cube function.

```
from joblib import Parallel, delayed
```

```
def cube(x):
    return x*x*x


Parallel(n_jobs=-1)(delayed(cube)(i) for i in range(10))
```

In the second example, we decorate the cube function using delayed, and using the more familiar @ syntax above the function definition and hence make the generator expression cleaner.

```
from joblib import Parallel, delayed


@delayed
def cube(x):
    return x*x*x


Parallel(n_jobs=-1)(cube(i) for i in range(10))
```

In the third example, we use the decorated cube function and produce an explicit generator expression. We then feed this generator expression to the Parallel class.

```
from joblib import Parallel, delayed


@delayed
def cube(x):
    return x*x*x


gen = (cube(i) for i in range(10))


Parallel(n_jobs=-1)(gen)
```

These three mechanisms produce the same result but the authors find the third method more readable.

In the last example, we will discuss a more realistic parallelization case. In this example, we will perform sigmoid correction that we discussed in Chapter 5. The sigmoid function takes the file name as input, reads the image using OpenCV, then performs sigmoid correction and stores the corrected image as a file. The function is decorated with @delayed so that it can be run in parallel. The generator expression (called gen in the example) accepts a list of file names, iterates over them and calls the sigmoid function for each image. When the code is executing, open the 'process monitor' for your operating system and you will notice multiple Python processes running.

```python
import os
import cv2
from skimage.exposure import adjust_sigmoid
from joblib import Parallel, delayed

@delayed
def sigmoid(folder, file_name):
    path = os.path.join(folder, file_name)
    img = cv2.imread(path)
    img1 = adjust_sigmoid(img, gain=15)
    output_path = os.path.join(folder,
        'sigmoid_'+file_name)
    cv2.imwrite(output_path, img1)


folder = 'input'
file_names = ['angiogram1.png', 'sem2.png',
    'hequalization_input.png']
gen = (sigmoid(folder, file_name) for file_name in
    file_names)
Parallel(n_jobs=-1)(gen)
print("Processing completed.")
```

Appendix B

Parallel Programming using MPI4Py

B.1   Introduction to MPI

Message Passing Interface (MPI) is a system designed for programming parallel computers. It defines a library of routines that can be programmed using Fortran or C and is supported by most hardware vendors. There are popular MPI versions both free and commercially available for use. MPI version 1 was released in 1994. The current version is MPI2. This appendix serves as a brief introduction to parallel programming using Python and MPI. Interested readers are encouraged to check the MPI4Py documentation [MPI20] and books on MPI [GLS99] and [Pac11] for more details.

MPI is useful on distributed memory systems and also on a shared memory system. The distributed memory system consists of a group of nodes (containing one or more processors) connected using a high-speed network. Each node is an independent entity and can communicate with other nodes using MPI. The memory cannot be shared across nodes, i.e., the memory location in one node is not accessible by a process running in another node. The shared memory system consists of a group of nodes that can access the same memory location from all nodes. Shared memory systems are easier to program using OpenMP, thread programming, MPI, etc., as they can be imagined as one large desktop. Distributed memory systems need MPI for node-to-node communication and can also be programmed using OpenMP or thread-based programming for within-node computation.

There is a large amount of literature available in print as well as online that teaches MPI and OpenMP programming [Ope20b]. Since this is about Python programming, we limit the scope of this section to programming MPI using Python. We will discuss one of the MPI wrappers for Python called MPI4Py. Before we begin the discussion on MPI4Py, we will explain the need for MPI in image processing computation.

B.2   Need for MPI in Python Image Processing

Image acquisition results in the collection of billions of voxels of 3D data. Analyzing these data serially by reading an image, processing it and then reading the next one will result in long computational time. It will cause a bottleneck, especially considering that most imaging systems are closer to real-time imaging. Hence it is critical to process the images in parallel. Consider an image processing operation that takes 10 minutes to process on one CPU core. If there are 100 images to be processed, the total computation time would be 1000 minutes. Instead, if the 100 images are fed to 100 different CPU cores, the images can be processed in 10 minutes, as all images are being processed at the same time. This results in a speedup of 100X. The image processing can be completed in minutes or hours instead of days or weeks. Also, many of the image processing operations such as filtering or segmentation can easily be parallelized. Hence, when one node is computing on one image, the second node can compute on a different image without the need for communication between the two nodes. Most educational and commercial institutions have either built or purchased supercomputers or clusters. Python along with MPI4Py can be used to run image processing computation faster on these systems.

B.3   Introduction to MPI4Py

MPI4Py is a Python binding built on top of MPI versions 1 and 2. It supports point-to-point communication and collective communication of Python objects. We will discuss these communications in detail. The Python objects that can be communicated need to be picklable, i.e., the Python objects can be saved using Python's pickle or cPickle modules or a numpy array.

The two modes of programming MPI are single instruction multiple data (SIMD) and single program multiple data (SPMD). In SIMD programming, the same instruction runs on each node but with different data. An image processing example of SIMD processing would be performing a filtering operation by dividing the image into sub-images and writing the result to one image file for each process. In such a case, the same instruction, filtering, is performed on each of the sub-images on different nodes. In SPMD programming, a single program containing multiple instructions runs on different nodes with different data. An example would be a different filtering operation in which the image is divided into subdivisions and filtered, but instead of writing the results to a file, one of the nodes collects the filtered images and arranges them before they are saved. In this case, most of the nodes perform the same operation of filtering, while one of the nodes performs an extra operation of collecting the output from the other nodes. Generally, SPMD operations are more common than SIMD operations. We will discuss SPMD-based programming here.

An MPI program is constructed such that the same program runs on each node. To change the behavior of the program for a specific node, a test can be made for the rank of that node (also called the node number) and alternate or additional instructions for that node alone can be provided.

B.4   Communicator

The communicator binds groups of processes in one MPI session. In its simplest form, an MPI program needs to have at least one communicator. In the following example, we use a communicator to obtain the size and rank of a given MPI program. The first step is to import MPI from MPI4Py. Then the size and rank can be obtained using the Get_size() and Get_rank() Python functions.

```
from mpi4py import MPI
import sys

size = MPI.COMM_WORLD.Get_size()
rank = MPI.COMM_WORLD.Get_rank()

print("Process %d among %d"%(rank, size))
```

This Python program may be run at the command-line. Typically, in a supercomputer setting, it is submitted as a job such as a portable batch system (PBS) job. An example of such a program is shown below. In the second line of the program, the number of nodes is specified using nodes, the number of processors per node using ppn, the amount of memory per processor using pmem, and the time for which the program needs to be executed using walltime. The walltime in this example is 10 minutes. In the third line of the program, the directory where the Python program and other files are located is specified. The program can be saved as a text file under the name "run.pbs." The name is arbitrary and can be replaced with any other valid file name for a text file.

```
#!/bin/bash
#PBS -l nodes=1:ppn=8,pmem=1750mb,walltime=00:10:00
```

```
cd $PBS_O_WORKDIR
module load python-epd
module load gcc ompi/gnu
mpirun -np 8 python firstmpi.py
```

The PBS script must be submitted using the command "qsub run.pbs." The queuing system completes its tasks and outputs two files: an error file containing any error messages generated during the program execution and an output file containing the content of command line output from the program. In the next few examples, the same PBS script will be used for execution with a change in the name of the Python file.

B.5 Communication

One of the important tasks of MPI is to allow communication between two different ranks or nodes as evidenced by its name "Message Passing Interface." There are many modes of communication. The most common are point-to-point and collective communication. Communication in the case of MPI generally involves transfer of data between different ranks. MPI4Py allows transfer of any picklable Python objects or numpy arrays.

B.5.1 Point-to-Point Communication

Point-to-point communication involves passing messages or data between only two different MPI ranks or nodes. One of these ranks sends the data while the other receives it.

There are different types of send and receive functions in MPI4Py. They are:

- Blocking communication

- Nonblocking communication

- Persistent communication

In blocking communication, MPI4Py blocks the rank until the data transfer between the ranks is completed and the rank can be safely returned to the main program. Thus, no computation can be performed on the rank until the communication is complete. This mode is inefficient as the ranks are idle during data transfer. The commonly used functions for blocking communication in MPI4Py are `send()`, `recv()`, `Send()`, `Recv()` etc.

In nonblocking communication, the node that is transferring does not wait for the data transfer to be completed before it begins processing the next instruction. In nonblocking communication, a test is executed at the end of data transfer to ensure its success while in blocking communication, the test is the completion of data transfer. The commonly used functions for nonblocking communication in MPI4Py are `isend()`, `irecv()`, `Isend()`, `Irecv()`, etc.

In some cases, the communication needs to be kept open between pairs of ranks. In such cases, persistent communication is used. It is a subset of nonblocking communication that can be kept open. It reduces the overhead in creating and closing communication if a nonblocking communication is used instead. The commonly used functions for point-to-point communication in MPI4Py are `Send_init()` and `Recv_init()`.

The following program is an example of blocking communication. The rank 0 creates a picklable Python dictionary called data that contains two key-value pairs. It then sends the "data" to the second rank using `send` function. The destination for this data is indicated in the `dest` parameter. Rank 1 (under the elif statement), receives the "data" using the `recv` function. The source parameter indicates that the data needs to be received from the rank 0.

```
from mpi4py import MPI
comm = MPI.COMM_WORLD
```

```
rank = comm.Get_rank()
if rank == 0:
    data = 'a': 7, 'b': 3.14
    comm.send(data, dest=1, tag=11)
    print("Message sent, data is: ", data)
elif rank == 1:
    data = comm.recv(source=0, tag=11)
    print("Message Received, data is: ", data)
```

B.5.2 Collective Communication

Collective communication allows transmission of data between multiple ranks simultaneously. This communication is a blocking communication. A few scenarios in which can be used are:

- "Broadcast" data to all ranks.

- "Scatter" a chunk of data to different ranks.

- "Gather" data from all ranks.

- "Reduce" data from all ranks and perform mathematical operations.

In broadcast communication, the same data is copied to all the ranks. It is used to distribute an array or object that will be used by all the ranks. For example, a Python tuple can be distributed to the various ranks as data that can be used for computation.

In the scatter method, the data is broken into multiple chunks and each of these chunks is transferred to different ranks. This method can be used for breaking (say) an image into multiple parts and transferring the parts to different ranks. The ranks can then perform the same operation on the different sub-images.

In the gather method, the data from different ranks are aggregated and moved to one of the ranks. A variation of the gather method is the "allgather" method. This method collects the data from different ranks and places them in all the ranks.

In the reduce method, the data from different ranks are aggregated and placed in one of the ranks after performing reduction operations such as summation, multiplication, etc. A variation of the reduce method is the "allreduce" method. This method collects the data from different ranks, performs reduction operations and places the result in all the ranks.

The program below uses broadcast communication to pass a 3-by-3 numpy array to all ranks. The numpy array containing all ones except for the central element is created in rank 0 and is broadcast using the bcast function.

```python
from mpi4py import MPI
import numpy

comm = MPI.COMM_WORLD
rank = comm.Get_rank()

if rank == 0:
    data = numpy.ones((3,3))
    data[1,1] = 3.0
else:
    pass
data = comm.bcast(data, root=0)
print("rank = ", rank)
print("data = ", data)
```

B.6　Calculating the Value of PI

The following program combines the elements of MPI that we have illustrated so far. The various MPI programming principles that will be used in this example are the MPI barrier, MPI collective communication, and specifically MPI reduce, in addition to MPI ranks.

The program calculates the value of PI using the Gregory-Leibniz series. A serial version of the program was discussed in Chapter 2, "Computing using Python modules." The program execution begins with the line "if __name__." The rank and size of the program are first obtained. The total number of terms is divided across the various ranks, so that each rank receives the same number of terms. Thus, if the program has 10 ranks and the total number of terms is 1 million, each rank will compute 100,000 terms. Once the number of terms is calculated, the "calc_partial_pi" function is called. This function calculates the "partial pi" value for each rank and stores it in the variable "partialval." The MPI barrier function is called to ensure that all the ranks have completed their computation before the next line, namely the comm.reduce() function, is executed to sum the values from various ranks and store it in the variable "finalval." Finally, the first rank prints the value of pi, namely the content of finalval.

```
from mpi4py import MPI
import sys
import numpy as np
import time


def calc_partial_pi(rank, noofterms):
    start = rank*noofterms*2+1
    lastterm = start+(noofterms-1)*2
    denominator = np.linspace(start, lastterm, noofterms)
    numerator = np.ones(noofterms)
```

```python
    for i in range(0, noofterms):
        numerator[i] = pow(-1, i+noofterms*rank)

    # Find the ratio and sum all the fractions
    # to obtain pi value
    partialval = sum(numerator/denominator)*4.0
    return partialval

if __name__ == '__main__':
    comm = MPI.COMM_WORLD
    rank = comm.Get_rank()
    size = MPI.COMM_WORLD.Get_size()
    totalnoterms = 1000000
    noofterms = totalnoterms/size

    partialval = calc_partial_pi(rank, noofterms)
    comm.Barrier()
    finalval = comm.reduce(partialval, op=MPI.SUM, root=0)
    if rank == 0:
        print("The final value of pi is ", finalval)
```

Appendix C

Introduction to ImageJ

C.1 Introduction

In all our discussions, we have used Python for image processing. There are many circumstances where it will be helpful to view the image so that it will be easy to prototype the algorithm that needs to be written in Python for processing. There are many such software programs, the most popular and powerful being ImageJ. This appendix serves as an introduction to ImageJ. Interested readers are encouraged to check the ImageJ documentation for more details at their website [Ins20].

ImageJ is a Java-based image processing software. Its popularity is due to the fact that it has an open architecture that can be extended by using Java and macros. Due to its open nature, there are many plug-ins written by scientists and experts that are available for free.

ImageJ can read and write most image formats and also specialized formats like DICOM, etc., similar to Python. Due to its ability to read and write images from many formats, ImageJ is popular in various fields of science. It is used for processing radiological images, microscope images, multi-modality images, etc.

ImageJ is available on most common operating systems such as Microsoft Windows, MacOSX and Linux.

C.2   ImageJ Primer

ImageJ can be installed by following the instructions at `http://rsb.info.nih.gov/ij/download.html`. Depending on the operating system, the methods for running ImageJ can vary. The instructions are available at the site listed above. Since ImageJ is written using Java, the interface looks the same across all operating systems, making it easier to transition from one operating system to another. Figure C.1 shows ImageJ on MacOSX.

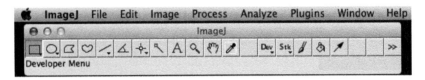

FIGURE C.1: ImageJ main screen

The files can be opened by using the File→Open menu. An example of this file is shown in Figure C.2. The 3D volume data that are stored as a series of 2D slice files can be opened using the File→Import→Image Sequence... menu.

In Chapter 3, "Image and its Properties," we discussed the basics of window and level. The window and level can be adjusted for the image in Figure C.2. They can be accessed using the Image→Adjust→Window/Level menu and they can be adjusted by using the sliders shown in Figure C.3.

We have previously discussed various image processing techniques like filtering, segmentation, etc. Such operations can also be performed using ImageJ using the Process menu. For example, the method for applying a median filter on the image is shown in Figure C.4.

Statistical information such as the histogram, mean, median, etc., of an image can be obtained using the Analyze menu. Figure C.5 demonstrates the method for obtaining the histogram using the Analyze menu.

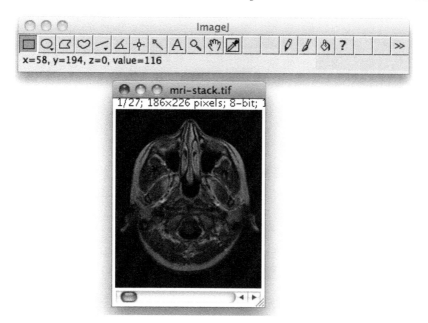

FIGURE C.2: ImageJ with an MRI image.

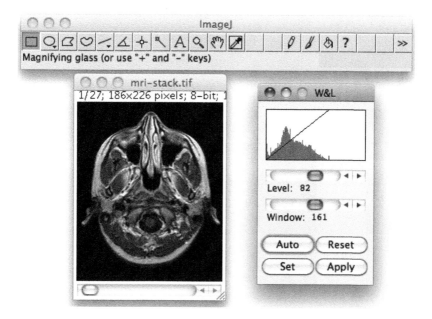

FIGURE C.3: Adjusting window or level on an MRI image.

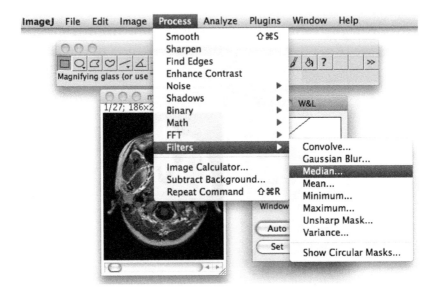

FIGURE C.4: Performing median filter.

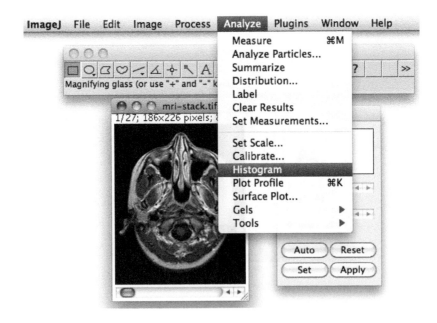

FIGURE C.5: Obtaining histogram of the image.

Appendix D

MATLAB® and Numpy Functions

D.1 Introduction

This appendix serves programmers who are migrating from MAT-LAB to Python and interested in converting their MATLAB scripts to equivalent Python programs using numpy.

MATLAB [Mat20b] is a popular commercial software that is widely used to perform computation in various fields of science including image processing. Both MATLAB and Python are interpreted languages. They both are dynamic typed, i.e., variables do not have to be declared before they are used. They both allow fast programming.

Numpy is similar in design to MATLAB in that they both operate on matrices. Because of their similarity we can find an equivalent function in MATLAB for a specific task in numpy and vice versa. The following table lists MATLAB functions and their equivalent numpy function. The first column has the numpy function, the second column contains the equivalent MATLAB function, and the last column gives the description of the function. A more extensive table can be found at [Sci20b].

Numpy Function	MATLAB Equivalent	Function Description
$a[a < 10] = 0$	$a(a < 10) = 0$	Elements in a with value less than 10 are replaced with zeros.

Numpy Function	MATLAB Equivalent	Function Description
$dot(a, b)$	$a * b$	Matrix multiplication.
$a * b$	$a. * b$	Element-by-element multiplication.
$a[-1]$	$a(end)$	Access the last element in the row matrix a.
$a[1, 5]$	$a(2, 6)$	Access elements in columns 2 and 6 in a.
$a[3]$ or $a[3 :]$	$a[4]$	Consider entire 4th row of a.
$a[0 : 3]$ or $a[: 3]$ or $a[0 : 3, :]$	$a(1 : 3, :)$	Access first three rows of a. In Python the last index is not included in the limits.
$a[-6 :]$	a(end-5:end,:)	Access the last six rows of a.
$a[0 : 5][:, 6 : 11]$	$a(1 : 5, 7 : 11)$	Access rows 1 to 5 and columns 7 to 11 in a.
$a[:: -1, :]$	$a(end : -1 : 1, :)$ or $flipud(a)$	Access rows in a in reverse order.

Numpy Function	MATLAB Equivalent	Function Description
$zeros((5,4))$	$zeros(5,4)$	Array of size 5-by-4 of zeros is created. The inner parentheses are used because the size of the matrix has to be passed as a tuple.
$a[r[: len(a), 0]]$	$a([1:end1],:)$	A copy of the first row will be appended at the end of matrix a.
$linspace(1,2,5)$	$linspace(1,2,5)$	Five equally spaced samples between and including 1 and 2 are created.
$mgrid[0:10.,0:8.]$	$[x,y] = meshgrid(0:10,0:8)$	Creates a 2D array with x-values ranging from [0,10] and y-values ranging from [0,8].
$shape(a)$ or $a.shape$	$size(a)$	Gives the size of a.
$tile(a,(m,n))$	$repmat(a,m,n)$	Creates m-by-n copies of a.
$a.max()$	$max(max(a))$	Output is the maximum value in the 2D array a.
$a.transpose()$ or $a.T$	a'	Transpose of a.
$a.conj().transpose()$ or $a.conj().T$	a'	Conjugate transpose of a.
$linalg.matrix\_rank(a)$	$rank(a)$	Rank of a matrix a.

Numpy Function	MATLAB Equivalent	Function Description
$linalg.inv(a)$	$inv(a)$	Inverse of square matrix a.
$linalg.solve(a, b)$ if a is a square matrix or $linalg.lstsq(a, b)$ otherwise	a/b	Solve for x in $ax = b$.
$concatenate((a, b), 1)$ or $hstack((a, b))$ or $column\_stack((a, b))$	$[a\ b]$	Concatenate columns of a and b along the horizontal direction.
$vstack((a, b))$ or $row\_stack((a, b))$	$[a; b]$	Concatenate columns of a and b along the vertical direction.

Bibliography

[ABC⁺16] Martin Abadi, Paul Barham, Jianmin Chen, Zhifeng Chen, Andy Davis, Jeffrey Dean, Matthieu Devin, Sanjay Ghemawat, Geoffrey Irving, Michael Isard, Manjunath Kudlur, Josh Levenberg, Rajat Monga, Sherry Moore, Derek G. Murray, Benoit Steiner, Paul Tucker, Vijay Vasudevan, Pete Warden, Martin Wicke, Yuan Yu, and Xiaoqiang Zheng. Tensorflow: A system for large-scale machine learning. In *12th USENIX Symposium on Operating Systems Design and Implementation (OSDI 16)*, pages 265–283, 2016.

[AFFV98] Koen L. Vinc Alejandro F. Frangi, Wiro J. Niessen and Max A. Viergever. Multiscale vessel enhancement filtering. *Medical Image Computing and Computer-Assisted Intervention (MICCAI) Lecture Notes in Computer Science*, 1998.

[Ana20a] Continuum Analytics. Anaconda. www.anaconda.com, 2020. Accessed on 14 January 2020.

[Ana20b] Continuum Analytics. conda. http://docs.conda.io, 2020. Accessed on 14 January 2020.

[Bea09] D.M. Beazley. *Python: Essential Reference*. Addison-Wesley Professional, Boston, MA, 2009.

[Bir11] W. Birkfellner. *Applied Medical Image Processing: A Basic Course*. Taylor & Francis, Boca Raton, FL, 2011.

[BK04] J. Barrett and N. Keat. Artifacts in CT: Recognition and avoidance. *Radiographics*, 24(6):1679–1691, 2004.

[BR98] J.J. Bozzola and L.D. Russell. *Electron Microscopy, 2nd ed.* Jones & Bartlett, Burlington, MA, 1998.

[Bra78] R.N. Bracewell. *Fourier Transform and its Applications.* McGraw-Hill, New York, NY, 1978.

[Bra99] R.N. Bracewell. *The Impulse Symbol.* McGraw-Hill, New York, NY, 1999.

[Bre12] E. Bressert. *SciPy and NumPy.* O'Reilly Media, Sebastopol, CA, 2012.

[BS13] F.J. Blanco-Silva. *Learning SciPy for Numerical and Scientific Computing.* Packt Publishing, Birmingham, England, 2013.

[Bus88] S.C. Bushong. *Magnetic Resonance Imaging.* CV Mosby, St. Louis, MO, 1988.

[Bus00] S. Bushong. *Computed Tomography.* Essentials of medical imaging series. McGraw-Hill Education, 2000.

[C+20] François Chollet et al. Keras. `https://keras.io`, 2020. Accessed on 22 Jan 2020.

[Can86] J. Canny. A computational approach to edge detection. *IEEE Transactions on Pattern Analysis and Machine Intelligence*, 8(6):679–698, 1986.

[CDM84a] T.S. Curry, J.E. Dowdey, and R.C. Murray. *Introduction to the Physics of Diagnostic Radiology.* Lea and Febiger, Philadelphia, PA, 1984.

[CDM84b] T.S. Curry, J.E. Dowdey, and R.C. Murry. *Christensen's Introduction to Physics of Diagnostic Radiology.* Lippincott Williams and Wilkins, Philadelphia, PA, 1984.

[CMSJ05] Y. Cho, D.J. Moseley, J.H. Siewerdsen, and D.A. Jaffray. Accurate technique for complete geometric calibration of cone-beam computed tomography systems. *Medical Physics*, 32:968–983, 2005.

[CT65] J.W. Cooley and J.W. Tukey. An algorithm for the machine calculation of complex Fourier series. *Mathematics of Computation*, 19:297–301, 1965.

[CV99] Tony Chan and Luminita Vese. An active contour model without edges. *Scale-Space Theories in Computer Vision*, pages 141–151, 1999.

[Dim12] C.A. Dimarzio. *Optics for engineers*. CRC Press, Boca Raton, FL, 2012.

[DKJ06] D. Dowsett, P.A. Kenny, and R.E. Johnston. *The Physics of Diagnostic Imaging, 2nd ed.* CRC Press, Boca Raton, FL, 2006.

[Dom15] Pedro Domingos. *The Master Algorithm: How the Quest for the Ultimate Learning Machine Will Remake Our World*. Basic Books, 2015.

[Dou92] E.R. Dougherty. *Introduction to Morphological Image Processing*. SPIE International Society for Optical Engineering, 1992.

[DR03] M.J. Dykstra and L.E. Reuss. *Biological Electron Microscopy: Theory, Techniques, and Troubleshooting*. Kluwer Academic/Plenum Publishers, Dordrecht, The Netherlands, 2003.

[ED82] J. C. Elliott and S. D. Dover. X-ray microtomography. *Journal of Microscopy*, 126(2):211–213, 1982.

[Eva10] L.C. Evans. *Partial Differential Equations, 2nd ed.* American Mathematical Society, 2010.

[FDK84]   L. Feldkamp, L. Davis, and J. Kress. Practical cone beam algorithm. *Journal of the Optical Society of America,* A6:612–619, 1984.

[FH00]   R. Fahrig and D.W. Holdsworth. Three-dimensional computed tomographic reconstruction using a c-arm mounted xrii: Image-based correction of gantry motion nonidealities. *Medical Physics,* 27(1):30–38, 2000.

[Fuk80]   Kunihiko Fukushima. Neocognitron: A self-organizing neural network model for a mechanism of pattern recognition unaffected by shift in position. *Biological Cybernetics,* 36(4):193–202, Apr 1980.

[GBC16]   Ian Goodfellow, Yoshua Bengio, and Aaron Courville. *Deep Learning.* MIT Press, 2016. `http://www.deeplearningbook.org`.

[GLS99]   W. Gropp, E.L. Lusk, and A. Skjellum. *Using MPI, 2nd ed.* The MIT Press, Boston, MA, 1999.

[Gol03]   J. Goldstein. *Scanning Electron Microscopy and X-ray Microanalysis,* volume v. 1. Kluwer Academic/Plenum Publishers, Dordrecht, The Netherlands, 2003.

[Gro17]   Aurlien Gron. *Hands-On Machine Learning with Scikit-Learn and TensorFlow: Concepts, Tools, and Techniques to Build Intelligent Systems.* O?Reilly Media, Inc., 1st edition, 2017.

[GT01]   D. Gilbarg and N.S. Trudinger. *Elliptic Partial Differential Equations.* Springer, New York, NY, 2001.

[GWE09]   R.C. Gonzalez, R.E. Woods, and S.L. Eddins. *Digital image processing using MATLAB®, 2nd ed.* Gatesmark Publishing, TN, 2009.

[Haj99] A.N. Hajibagheri. *Electron Microscopy: Methods and Protocols.* Humana Press, New York, NY, 1999.

[Hay00] A. Hayat. *Principles and Techniques of Electron Microscopy: Biological Applications.* Cambridge University Press, Cambridge, England, 2000.

[HBS13] C.L.L. Hendriks, G. Borgefors, and R. Strand. *Mathematical Morphology and Its Applications to Signal and Image Processing.* Springer, New York, NY, 2013.

[Hen83] W.R. Hendee. *The Physical Principles of Computed Tomography.* Little, Brown library of radiology. Little Brown, New York, NY, 1983.

[Het10] M.L. Hetland. *Python Algorithms: Mastering Basic Algorithms in the Python Language.* Apress, New York, NY, 2010.

[HK93] S.L. Fleglerand J.W. Heckman and K.L. Klomparens. *Scanning and Transmission Electron Microscopy: An Introduction.* Oxford University Press, Oxford, England, 1993.

[HL93] B. Herman and J.J. Lemasters. *Optical microscopy: Emerging Methods and Applications.* Academic Press, Waltham, MA, 1993.

[Hor95] A.L. Horowitz. *MRI Physics for Radiologists: A Visual Approach.* Springer-Verlag, New York, NY, 1995.

[HS88] Chris Harris and Mike Stephens. A combined corner and edge detector. *In Proc. of Fourth Alvey Vision Conference*, pages 147–151, 1988.

[Hsi03] J. Hsieh. *Computed Tomography: Principles, Design, Artifacts, and Recent Advances.* SPIE, 2003.

[HWJ98]   L. Hong, Y. Wan, and A. Jain. Fingerprint image enhancement: algorithm and performance evaluation. *IEEE Transactions on Pattern Analysis and Machine Intelligence*, 20(8):777–789, 1998.

[Idr12]   I. Idris. *NumPy Cookbook*. Packt Publishing, Birmingham, England, 2012.

[IK87]   J. Illingworth and J. Kittler. The adaptive Hough transform. *IEEE Transactions on Pattern Analysis and Machine Intelligence*, 9(5):690–698, 1987.

[IK88]   J. Illingworth and J. Kittler. A survey of the Hough transform. *Computer Vision, Graphics, and Image Processing*, 44(1):87–116, 1988.

[Ins20]   National Health Institute. ImageJ documentation. `http://imagej.nih.gov/ij/docs/guide/`, 2020. Accessed on 21 Jan 2020.

[JKRL09]   K. Jarrett, K. Kavukcuoglu, M. Ranzato, and Y. LeCun. What is the best multi-stage architecture for object recognition? *2009 IEEE 12th International Conference on Computer Vision*, pages 2146–2153, Sep. 2009.

[Job20]   Joblib. `https://joblib.readthedocs.io/`, 2020. Accessed on 21 January 2020.

[JS78]   P.M. Joseph and R.D. Spital. A method for correcting bone induced artifacts in computed tomography scanners. *Journal of Computer Assisted Tomography*, 2:100–108, 1978.

[Kal00]   W. Kalender. *Computed Tomography: Fundamentals, System Technology, Image Quality, Applications*. Publicis MCD Verlag, 2000.

[KB46] E. Kohl and W. Burton. *The Electron Microscope; An Introduction to Its Fundamental Principles and Applications.* Reinhold, 1946.

[Key97] R.J. Keyse. *Introduction to Scanning Transmission Electron Microscopy.* Microscopy Handbooks. Bios Scientific Publishers, Oxford, England, 1997.

[KS88] A.C. Kak and M. Slaney. *Principles of Computerized Tomographic Imaging.* IEEE Press, New York, NY, 1988.

[Kuo07] J. Kuo. *Electron Microscopy: Methods and Protocols.* Methods in Molecular Biology. Humana Press, New York, NY, 2007.

[LBBH98] Y. Lecun, L. Bottou, Y. Bengio, and P. Haffner. Gradient-based learning applied to document recognition. *Proceedings of the IEEE*, 86(11):2278–2324, Nov 1998.

[LBD⁺89] Y. LeCun, B. Boser, J. S. Denker, D. Henderson, R. E. Howard, W. Hubbard, and L. D. Jackel. Backpropagation applied to handwritten zip code recognition. *Neural Computation*, 1(4):541–551, Dec 1989.

[LCB10] Yann LeCun, Corinna Cortes, and CJ Burges. Mnist handwritten digit database. *ATT Labs. Available: http://yann.lecun.com/exdb/mnist*, 2, 2010.

[Lew95] J.P. Lewis. Fast template matching. *Vision Interface*, 95:120–123, 1995.

[LK87] L.A. Love and R.A. Kruger. Scatter estimation for a digital radiographic system using convolution filtering. *Medical Physics*, 14(2):178–185, 1987.

[LLM86] H. Li, M.A. Lavin, and R.J. Le Master. Fast hough transform: A hierarchical approach. *Computer Vision, Graphics, and Image Processing*, 36(2-3):139–161, 1986.

[Lut06] M. Lutz. *Programming Python*. O'Reilly, Sebastopol, CA, 2006.

[Mac83] A. Macovski. *Medical Imaging Systems*. Prentice Hall, Upper Saddle River, NJ, 1983.

[Mar72] A. Martelli. Edge detection using heuristic search methods. *Computer Graphics and Image Processing*, 1(2):169–182, 1972.

[Mas20] et al Mason, D. L. `https://github.com/pydicom/pydicom`, 2020. Accessed on 22 Jan 2020.

[Mat20a] Materialise. Mimics™. `http://biomedical.materialise.com/mimics`, 2020. Accessed on 22 Jan 2020.

[Mat20b] Mathworks. Matlab®. `https://www.mathworks.com/`, 2020. Accessed on 21 Jan 2020.

[MB90] F. Meyer and S. Beucher. Morphological segmentation. *Journal of Visual Communication and Image Representation*, 1(1):21–46, 1990.

[McR03] D.W. McRobbie. *MRI from Picture to Proton*. Cambridge University Press, Cambridge, England, 2003.

[Mer10] J. Mertz. *Introduction to Optical Microscopy*. Roberts and Company, Greenwood Village, CO, 2010.

[Mey92] F. Meyer. Color image segmentation. *Proceedings of the International Conference on Image Processing and its Applications*, pages 303–306, 1992.

[Mey94] F. Meyer. Topographic distance and watershed lines. *Signal Processing*, 38:113–125, 1994.

[MH80] D. Marr and E. Hildreth. Theory of edge detection. *Proceedings of the Royal Society of London. Series B, Biological Sciences*, 207(1167):187–217, 1980.

[MPI20] MPI4Py.org. Mpi4py. `https://mpi4py.readthedocs.io/`, 2020. Accessed on 22 Jan 2020.

[MTH] Mark H. Beale Orlando De Jesus Martin T. Hagan, Howard B. Demuth. *Neural Network Design (2nd Edition)*. `http://hagan.okstate.edu/NNDesign.pdf`.

[MW98] J.A. Markisz and J.P. Whalen. *Principles of MRI: Selected Topics*. Appleton & Lange, East Norwalk, CT, 1998.

[NT10] L. Najman and H. Talbot. *Mathematical Morphology*. Wiley-ISTE, 2010.

[OFKR99] B. Ohnesorge, T. Flohr, and K. Klingenbeck-Regn. Efficient object scatter correction algorithm for third and fourth generation CT scanners. *European Radiology*, 9:563–569, 1999.

[Ope20a] OpenCV. `http://docs.opencv.org`, 2020. Accessed on 21 Jan 2020.

[Ope20b] OpenMP.org. OpenMP. `http://openmp.org/`, 2020. Accessed on 22 Jan 2020.

[OR89] S. Osher and L.I. Rudin. Feature-oriented image enhancement using shock filters. *SIAM Journal Numerical Analysis*, 27(4):919–940, 1989.

[Ots79] N. Otsu. A threshold selection method from gray level histograms. *IEEE Transactions on Systems, Man and Cybernetics*, 9(1):62–66, 1979.

[Pac11] P. Pacheco. *An Introduction to Parallel Programming*. Morgan Kaufmann, Burlington, MA, 2011.

[Par91] J.R. Parker. Gray level thresholding in badly illuminated images. *IEEE Transactions on Pattern Analysis and Machine Intelligence*, 13:813–819, 1991.

[PC18] Sridevi Pudipeddi and Ravi Chityala. *Essential Python*. Essential Education, 2018.

[PK81] S.K. Pal and R.A. King. Image enhancement using smoothing with fuzzy sets. *IEEE Transactions on Systems, Man, and Cybernetics*, 11(7):494–501, 1981.

[PK91] M. Petrou and J. Kittler. Optimal edge detectors for ramp edges. *IEEE Transactions on Pattern Analysis and Machine Intelligence*, 13(5):483–491, 1991.

[ppr20] Hubel Weisel Nobel prize press release. `https://www.nobelprize.org/prizes/medicine/1981/press-release/`, 2020. Accessed on 19 January 2020.

[Pre70] J.M.S. Prewitt. Object enhancement and extraction. *Picture Processing and Psychopictorics*, pages 75–149, 1970.

[R95] W. Röntgen. On a new kind of rays. *Würzburg Physical and Medical Society*, 137:132–141, 1895.

[RD06] Edward Rosten and Tom Drummond. Machine learning for high-speed corner detection. *ECCV*, 2006.

[RDLF05] K. Rogers, P. Dowswell, K. Lane, and L. Fearn. *The Usborne Complete Book of the Microscope: Internet Linked*. Complete Books. EDC Publishing, Tulsa, OK, 2005.

[Ren61] A. Renyi. On measures of entropy and information. *Proceedings of Fourth Berkeley Symposium on Mathematics Statistics and Probability*, pages 547–561, 1961.

[RHW86] David E. Rumelhart, Geoffrey E. Hinton, and Ronald J. Williams. Learning representations by back-propagating errors. *Nature*, 323(6088):533–536, 1986.

[Rob77] G.S. Robinson. Detection and coding of edges using directional masks. *Optical Engineering*, 16(6):166580–166580, 1977.

[Ror20] Chris Rorden. `https://people.cas.sc.edu/rorden/ezdicom/index.html`, 2020. Accessed on 22 Jan 2020.

[Rus11] J.C. Russ. *The Image Processing Handbook, 6th ed.* CRC Press, Boca Raton, FL, 2011.

[SAR20] Pixmeo SARL. `http://www.osirix-viewer.com/`, 2020. Accessed on 22 Jan 2020.

[Sch89] R.J. Schalkoff. *Digital Image Processing and Computer Vision*. Wiley, New York, 1989.

[Sch04] H.M. Schey. *Div, Grad, Curl, and All That, 4th ed.* W.W. Norton and Company, New York, NY, 2004.

[Sci20a] Scikits.org. Scikits. `http://scikit-image.org/docs/dev/api/api.html`, 2020. Accessed on 22 Jan 2020.

[Sci20b] SciPy.org. Numpy to MATLAB℗. `https://docs.scipy.org/doc/numpy/user/numpy-for-matlab-users.html`, 2020. Accessed on 21 Jan 2020.

[Sci20c] SciPy.org. Scipy. `http://docs.scipy.org/doc/scipy/reference`, 2020. Accessed on 22 Jan 2020.

[Sci20d] SciPy.org. Scipy ndimage. `http://docs.scipy.org/doc/scipy/reference/ndimage.html`, 2020. Accessed on 22 Jan 2020.

[Ser82] J. Serra. *Image analysis and mathematical morphology*. Academic Press, Waltham, MA, 1982.

[Sha48] C.E. Shannon. A mathematical theory of communication. *Bell System Technical Journal*, 27:379–423, 1948.

[Sha96] V.A. Shapiro. On the Hough transform of multi-level pictures. *Pattern Recognition*, 29(4):589–602, 1996.

[SHB+99] M. Sonka, V. Hlavac, R. Boyle, et al. *Image Processing, Analysis, and Machine Vision*. PWS, Pacific Grove, CA, 1999.

[Si20] Scikits-image.org. Scikits-image. `https://scikit-image.org/docs/dev/api/skimage.measure.html`, 2020. Accessed on 22 Jan 2020.

[Smi07] J.O. Smith. *Mathematics of Discrete Fourier Transform: With Audio Applications*. W3K, 2007.

[Soi04] P. Soille. *Morphological Image Analysis: Principles and Applications, 2nd ed.* Springer, New York, NY, 2004.

[SPK98] L. Shafarenko, H. Petrou, and J. Kittler. Histogram-based segmentation in a perceptually uniform color space. *IEEE Transactions on Image Processing*, 7(9):1354–1358, 1998.

[Spl10] R. Splinter. *Handbook of Physics in Medicine and Biology*. CRC Press, Boca Raton, FL, 2010.

[SS94] J. Serra and P. Soille. *Mathematical Morphology and Its Applications to Image Processing*. Springer, New York, NY, 1994.

[SS03] E. Stein and R. Shakarchi. *Fourier Analysis: An Introduction*. Princeton University Press, Princeton, NJ, 2003.

[SSW88] P.K. Sahoo, S. Soltani, and A.K.C. Wong. A survey of thresholding techniques. *Computer Vision, Graphics, and Image Processing*, 4(8):233–260, 1988.

[Vai09] S. Vaingast. *Beginning Python Visualization: Crafting Visual Transformation Scripts*. Apress, New York, NY, 2009.

[Wat97] I.M. Watt. *The Principles and Practice of Electron Microscopy.* Cambridge University Press, Cambridge, England, 1997.

[Wes09] C. Westbrook. *MRI at a Glance.* Wiley, New York, NY, 2009.

[WSOV96] G. Wang, D.L. Snyder, J.A. O'Sullivan, and M.W. Vannier. Iterative deblurring for CT metal artifact reduction. *IEEE Transactions on Medical Imaging*, 15:657–664, 1996.

[WSS01] Wes Wallace, Lutz H. Schaefer, and Jason R. Swedlow. A working persons guide to deconvolution in light microscopy. *BioTechniques Open Access.*, 31(5):1076–1097, 2001.

[XO93] L. Xu and E. Oja. Randomized Hough transform: Basic mechanisms, algorithms, and computational complexities. *Computer Vision, Graphics, and Image Processing*, 57(2):131–154, 1993.

[ZMP⁺02] W. Zhang, S. Mukhopadhyay, S.V. Pletnev, T.S. Baker, R.J. Kuhn, and M.G. Rossmann. Placement of the structural proteins in Sindbis virus. *Journal of Virology*, 76:11645–11658, 2002.

[Zui94] Karel Zuiderveld. Contrast limited adaptive histogram equalization. *Graphics gems IV*, pages 474–485, 1994.

Index

Printed and bound by CPI Group (UK) Ltd, Croydon, CR0 4YY

24/10/2024

01778304-0018